SCIENCE
AND
A GLOBAL ETHIC

SCIENCE
AND
A GLOBAL ETHIC

F.R.J. Williams

PILKINGTON PRESS

First Published 1994
By Pilkington Press Ltd
Yelvertoft Manor
Yelvertoft
Northampton NN6 7LF

© F.R.J. Williams 1994

All rights reserved.

No part of this publication may be reproduced, stored in a retrieval system, or transmitted in any form or by any means, electronic, mechanical, photocopying, recording or otherwise, without the prior permission of the publisher and copyright owner.

The right of F.R.J. Williams to be identified as author of this work has been asserted generally in accordance with the Copyright, Designs and Patents Act 1988.

ISBN 1 899044 04 3

Produced, Designed and Typeset by
A.H. Jolly (Editorial) Ltd
Yelveroft Manor
Northamptonshire NN6 7LF

Printed in Great Britain

CONTENTS

7   INTRODUCTION BY DR T. KAWAI

9   FOREWORD BY PROFESSOR DUNCAN DERRETT DD

15 PREFACE

PART I. THE EVOLVING WORLD

19 Religious Implications of Roy Peacock's book, *A Brief History of Eternity*

21 Revelation and the Mysterious

23 The Failure of Monotheistic Religions

24 A Philosophical View of Man and Evolution

27 The Faculty of Self-awareness

28 Evolution in Perspective

29 Homo sapiens as part of the Evolutionary Hierarchy

31 How Free is Man's Will?

33 Evolutionary Change: Higher or merely Different?

34 The Importance of Knowledge and Love

PART II. THE RELIGIOUS DIMENSION

39 Differing Attitudes to Religious Faith

41 The Two Main Streams of Living Faith: East and West

42 The Western 'Prophetic' Religions

43 Eastern Religion

44   The Relevance of Mythology

47   Fundamentalism

48   Religious Pluralism and the Process of Change

50   Convergence of Eastern and Western Ideas

51   Material Idealism: Engels and Teilhard de Chardin

55   The New Age and Green Movements

57   A Religiously Plural World

59   The Dilemma of Christianity

PART III. AN HOLISTIC PHILOSOPHY

67   The Unifying Potential of Global Values

71   Global Values and the Established Religions

76   A World Ethic

80   Global Values and the Environment

83   An Holistic Philosophy in Perspective

86   Conclusion and Summary

91   BIBLIOGRAPHY

INTRODUCTION

The necessity to promote exchange between East and West

In today's democratic and industrialized society, the attempt to articulate and define the meaning of such complex issues as family life, social cohesion and religion, may appear to be a futile exercise. In the pluralism of the western world, religious values are diverse and ambiguous. In Japan, religion has remained undefined. None the less, it could be affirmed that the essence of all religions must be the same throughout the world as they all result from basic human needs which are as innate as they are universal.

Amidst the expansion and progress of science, it is increasingly difficult for individuals within any particular society to consider objectively its inherent values. Basic religious beliefs are taken for granted by different cultural societies, while its members pursue individualistic and material ends. These pursuits threaten the destruction of our natural resources and leave mankind with the fear of extinction from atomic warheads and other means of mass destruction created by scientific advances.

The world may have become smaller because of jet travel and the pervasiveness of the mass media, but true communication by the bridging of differing cultural values still eludes us. Yet communication and understanding is now imperative if peoples are to establish a common global interest to protect the limited natural resources required for the survival of the human race.

It is the wish of the publishers to seek to promote dialogue between different value systems, especially those of the English and Japanese peoples which represent the essential differences between East and West in the developed industrial and democratic world.

Useful comparisons can be drawn between attitudes towards life, family, science, art, religion, money and social cohesion.. Conflicts will occur in all these areas if people tend to take their own values for granted. It is essential to reveal the superficial conflicting differences and to create harmony by looking to the future and its challenges rather than by seeking to impose our values on each other. By a sympathetic analysis of the cultural differences between East and West, we trust a less divisive and mutually acceptable value system will be forged, which will help to resolve the conflicts of needs and hopes of the developed and underdeveloped nations of the world.

Dr T. Kawai
Far East Director
Pilkington Foundation

FOREWORD
by J.D.M. Derrett

The author of the present work argues that it is crucial for the world's future that humankind should work to bring into being a world ethic or global moral values. He is concerned about the sentiments of men and women of good will, and also about decision-making, and about action. Can a common global ethic evolve, as a mental achievement, from the existing principles on which human behaviour has been based in widely different types of society, almost simultaneously? It will be remembered that those societies have lived under regimes and thought-patterns which have stood the test of time, until, that is, our instant dilemmas find all of them simultaneously wanting.

It does not follow that arrangements made by *states*, as distinguished from societies, will meet such a case. Perhaps it may be helpful to try to take stock, briefly, of the sources of a sense of duty which have come into vogue and been recorded hitherto. Broadly drawn patterns have obtained over thousands of years, and history explains much. The behaviour of South American societies is better understood, for example, if one is well briefed in Roman Law. Mr F.R.J. Williams is not concerned to proceed historically. He rightly applies his mind to the whole question of global moral values, as every active enquirer will, *rebus sic stantibus,* in the exact juncture of all our affairs. But to some extent response to the challenges which are now reaching a climax is confined by history, by the development of human nature in different environments. He takes for granted a screen or backdrop of the past. We are where we stand because of the paths we have trodden. And because we are many, the paths are many. It will not be inopportune to take a bird's-eye view of those paths, as they are and as they are perceived. In fact they are converging. Good will, as such, must not blind us to their original diversity. Let ethics and morals become never so well assimilated, the laws that obtain their legitimacy from them will be bound to differ in scope, style and wording, not to speak of efficacy. The historical arena in which these practical questions will be tackled prepares the mind for Mr Williams' essay.

The Roman jurist Gaius (2nd cent. AD) took, from the vantage point of Rome, a quick view over the laws of the world. Ever since Pliny the Elder, Rome had been equipped with at least one encyclopedia, and to Rome herself resorted men who had travelled over all known lands and seas. Gaius observed that though peoples differed they shared many ways.

> The laws of every people governed by statutes and customs are partly peculiar to itself, partly common to all mankind. The rules established by a given state for its own members are peculiar to itself, and are called *ius civile* (civil law); the rules constituted by natural

reason for all human beings are observed amongst all nations equally, and are called *ius gentium* (the law of peoples), for this law applies to all peoples. So the laws of the people of Rome are partly peculiar to itself and partly common to all mankind...

It is a serious question whether any peoples have ever shown an ability to act justly by the *light of nature*. Stoics in Gaius' time believed the (very rare) wise man might do so, and therefore that it was theoretically possible to comply with the *law of nature*. The appeal, if not nobility, of Stoic ethics was consistent with this. Unfortunately, on account of the extreme rarity of such wise men, no detailed *content* of a law of nature ever emerged, and it is said this was perceived by St Paul also. To his know-ledge there was no religion, and no system of philosophy then extant, which sustained any identifiable people in the practice of righteousness by the light of nature.[1]

Gaius was, perhaps, optimistic and superficial. Romans knew well that of the systems of law and custom, precept and practice, which then existed there were two, Hindu and Jewish, which not only claimed to be exhaustive and exclusive amongst those who professed the corresponding religions but, when expounded by their revered jurists and moralists, could also seriously be called inerrant and irrefutable. They did not agree with each other, save for the fundamentals of civilized living, and perhaps not always in every aspect of those: but they were alike in two ways of which Rome could not but be aware.

Firstly, the societies governed by the Torah or by the Vedic and post-Vedic scriptures respectively were 'collective', as opposed to non-collective, individualistic societies – such as Greece and Rome, who possessed no such 'scriptures', were fast becoming. Secondly, to each of those Asian nations laws, precepts and admonitions had ostensibly been handed down by divine beings, or sages wholly inspired by a spirit, as it were, from on high. This feature would be strange to Greeks, whose gods had a mild joint interest in morality (at most), and amongst those deities only Apollo had any (and a precarious) interest in honesty and good conscience. It would be stranger still to Romans, who did not regard law and morality as in any way proceeding from some voice external to Rome herself. Indeed they begot the *ius civile*, pronounced as cases required by an official and public-spirited judges in collaboration, a law based on observation and reason, and eventually expounded systematically by jurists with a devotion to reason, the results of which the world has envied ever since.

Gaius, one of these, was not prepared to notice the quaint arrangements of the Jews or of the Hindus, though both were enquired into by the restless minds of his contemporaries. The notable distinction between the collective and the individualistic society remains; for the idea that an individual can be a law to him- or herself, an idea we tend to deplore, finds far more scope in the latter than in the former; and as the world foments, day by day, the development of collective societies into their

---

1. A new view of Romans 2:14–16 propounded by John W. Martens, 'Romans 2.14–16: A Stoic reading,' *New Testament Studies* 40/1 (1994), 55–67. The non-Jews were able to divine what law they should live by through their use of *reason* (Romans 1:20–22), but there is no evidence they ever did it. This does not preclude a better use of reasoning powers hereafter.

opposite, we hark back with a natural nostalgia to the time (not entirely imaginary) when decision-making was a kind of fermenting, and no individuals or even groups offered themselves as innovators in respect of law and morals. Their 'originality', if known at all, was a direct threat to stability and to the comfort of the whole.

Slowly, as education opened minds and prosperity offered a variety of opportunities, individuals (however consciously conventional) began to hold their own needs in balance with those of the group, and the more independent-minded began to take note of where the society, even the greater society of humankind, circumscribed their ambitions. A monolithic scriptural authority can prove wholly intractable unless intelligently, and continually, updated with competent exegesis and a ready adaptation to the changing facts of life. Even Roman law itself, when actually codified under Justinian, reached a point of intellectual perfection only to outstrip the ability of human expositors to apply it in practice, as individuals and groups demanded a greater degree of flexibility than those intellectuals could, placed as they were, supply.

Meanwhile, many societies no less collective than ancient restrained their young by reference neither to scriptures, nor to codes (save for lists of highly generalized precepts), nor to expositors, nor even to nature (whose voice was muted when questions of moral tension arose), but to *ad hoc* pronouncements based on the sentiment of the society - in other words norms were customs rehearsed by society itself. They were not visualized as proceeding from *outside,* whether from a deity, or from a ruler or legislature. The ruler himself was subject to overriding considerations, or at least it was supposed that he was. In many places and at many periods rulers, pretending to legislate, merely made choices, for convenience's sake, from several equally successful customs in vigour in their realms. At length, in relatively modern times, a want of harmony between these inconsistent concepts of duty and right made itself felt, and a solution became more pressing, without a single authority being so placed as to provide one. The development of international agencies witnesses this gap in human ingenuity.

Gaius believed societies and peoples *shared* laws. He did not say that they accepted a common *source of law.* He never ventured beyond the surmise that reason should furnish a basic legal system - in which case, provided men and women agreed what was reasonable, a kind of 'common law' could develop (not unlike our present International Common Law), which all peoples might cheerfully obey. A laboratory at last did emerge in which these ideas not only could be tested but came up for testing. It was the continent duly called America.

All observations of Benjamin Franklin (born Boston, MA, in 1706) are worth study. From unprivileged beginnings his manifold qualities, as constitutional lawyer, philosopher, natural scientist, pamphleteer, statesman and diplomat, brought him to honour in the Old as well as the New World, and he participated in drafting the Declaration of Independence (1776). While still a youth, brought up as a dissenter from the Church of England, he felt obliged to forge a workable religion for himself, surrounded as he then was by Roman Catholics, by various mutually hostile sects of Protestantism, by rampant 'enthusiasts', and by the primeval tribal peoples that were the 'native Americans'. He settled for being a Deist. That was not an unfashionable decision in the 1720s, from which period the following reflections of his come:

> Revelation had indeed no weight with me, as such; but I entertained an opinion that, though certain actions might not be bad *because* they were forbidden by it, or good *because* it commanded them, yet probably those actions might be forbidden *because* they were bad for us, or commanded *because* they were beneficial to us, in their own natures, all the circumstances of things considered. And this persuasion, with the kind hand of Providence, or some guardian angel, or accidental favourable circumstances and situations, or all together, preserved me, through this dangerous time of youth, and the hazardous situations I was sometimes in among strangers, remote from the eye and advice of my father, without any wilful gross immorality or injustice, that might have been expected from my want of religion.[2]

It can hardly be an accident that, consonant with the temper of the times, which favoured the reinterpretation of various versions of Christianity as harmonious with Nature (Bishop Joseph Butler's *Analogy of Religion* appeared in 1736), the circumstances of the American Colonies favoured not merely freedom of religion, and the separation of religion from the state (see the First Amendment to the Constitution, 1791), but the gradual softening of the edges of controversy and jealousy as between the sects and faiths, and the eventual emergence of a common style of religious practice irrespective of the ancient traditions imported from the Old World, which remained still grievously divided, with its distinct revelations, prophets, intractabilities and antagonisms. For example, a citizen who is charitably inclined will now find any church a suitable medium for his instincts; the citizen is an American first and a member of a denomination second. It is of interest that Franklin himself, who seldom attended a place of worship, always responded generously to all applications for a subvention from whichever church approached him.

There can be little doubt that Revelation, in spite of its entire unsuitability to ground a morality, plays a wider role in religions which cater for societies closer to the collective past: it supplies the place of reason, whilst affording the unreasoning a ground for instantly adhering to a vague package of norms. In those societies, notably in the Far East, where Revelation plays a minimal role, reason still struggles to proffer some direction in the same quandaries that plague the remaining co-denizens of our planet, while the courage which the independent thinker, or preacher, requires is greater than the quandaries themselves can account for, where to preach concern for *foreign* peoples seems inexpressibly eccentric, and to love peace is to be called a hypocrite, as that great worker for peace, Leo Tolstoy, both was, and is.

---

2. *The Life of Benjamin Franklin Written by Himself,* World's Classics, ed., Oxford University Press, 1924 reprinted 1943, pp. 77–8.

Association is strength, as every society believes, but such pioneers as have questioned the direction of their collective must struggle to escape, if reform is not perceptible, an association cemented in ignorance, obscurantism, chauvinism and complacency.

So wherever it is pretended, as every child will claim, that decisive guidance comes from outside, the future is doomed to stagnation; while if society accepts no norms but those spun from its own belly, chaos, or the least developed media of social control may prevail. There must come a point at which the limits of the demands of the collective and the bounds necessary to the schemes of individuals are perceived to coincide. An overarching ethic may develop out of various traditions and express itself in various idioms. But at that point reason, based on careful observation, is bound to play its part, whether or not the actual reasoning is patent on the face of the record.[3]

Our world shrinks by the hour. Inhabitants of New Guinea already comment on the antics of middle-class 'whites'. Their reactions agree substantially with those of fellow-observers three-quarters of the globe away. Local norms, however they originate, will, as Gaius says, preserve their validity and their scope; but concerns held in common by peoples now able to debate their mutual interests as never before will eventually yield to joint discussion. Societal *as well as* environmental problems will come high on the agenda. The short-term ambitions of individuals or cliques will evoke similar reservations – and wherever, and in whatever respect, humankind is agreed action cannot lag far behind, societies leading, perhaps, where states will follow, as in international expedients familiar to us under the name of 'sanctions' and in other appropriate ways.

This very generation may find itself, despite all that evolution has done to forge our superficial appearances (not obliterating the fact which intrigues many, that we remain one species) and despite our ingrained prejudices, slowly emerging from behind and beyond mere municipal laws, and finding a meaningful common ethic, if only in respect of our 'one world', and in any case wholly irrespective of mythology, eschatology, cultural triumphalism and hysteria.[4] Mr Williams, it seems, has long visualized the making of this highly needful first step.

---

3. David Hume (1711-76) took the view that 'Reason is and ought only to be the slave of the passions, and can never pretend to any other office than to serve and obey them' (*A Treatise of Human Nature*, 11, iii, 3, ed. Selby-Bigge, Oxford, Clarendon Press, 1896, p. 415). This salutary warning does not mean that when one has perceived what one's motives are one cannot use reason to implement them, nor that reason may not be employed to discover what are our motives as distinct from our supposed motives.

4. 'Men botched and bungled by their inner conflicts resort to callous destruction of their fellow creatures in suicidal wars in order to protect the institutional structures of belief and practice that hold their society together' (Nolan P. Jacobson, *Buddhism: the Religion of Analysis*, London, Allen & Unwin, 1966, p. 22). I would say 'in order *ostensibly* to protect the institutional structures...' Men and women are very good at deceiving themselves as to what their actual motives are.

PREFACE

A cartoon appeared in the magazine *The New Yorker* some years ago depicting a meeting of important-looking directors of a large corporation with the caption: 'The question before the board, then, is whether or not to enter an altered state of consciousness.' The cartoon was poking fun at, but at the same time making a serious point about, the feeling among an increasing number of people that spiritual values are being swamped in our science-based, industrial era, and that a basic reappraisal is needed of the direction in which our civilization is heading. There is no doubt that the modern world is becoming increasingly materialistic in its outlook. Progress is defined in terms of economic growth and increasing consumption. The main driving forces in the more advanced economies are the quest for material advancement and short-term gain. This has succeeded in raising the standard of living in much of the world, but at the same time it has led to irresponsible exploitation of the world's resources. There is little or no commitment to ascertain whether these living standards can apply world-wide or whether they are sustainable for future generations. With the spectacular advances in science and technology, society is structured to demand and reward specialization, so that fragmentation and alienation have become the general rule. These attitudes are having a disastrous effect on the natural environment and tend to aggravate social and international problems, and if the world is to become a better place, or indeed not to deteriorate further, it is vital that these attitudes be transformed or replaced by a holistic perspective and a better understanding of what gives value to life on our planet.

In two books by professors working in the field of physical science, Stephen Hawking's *A Brief History of Time,* published in 1988, and a considered response to this by Roy Peacock in *A Brief History of Eternity,* published in 1989, the claim is made that in physics the search is on for 'the mind of God'. They arrive at different conclusions: but it seemed clear to me on reading them that scientific exploration of the physical mysteries of space and the universe, however interesting and fascinating in itself, is not relevant to the central problems confronting humankind in our modern multicultural world. It was this consideration that prompted me to take a fresh look at the way our civilization is heading, and at the basic questions which *Homo sapiens* has asked since earliest times: What is his place and function in the universe? And what gives meaning and value to life on earth?

This work thus represents a personal voyage of discovery. It now seems clear that our living planet is at a critical stage in its evolution, and it is increasingly urgent for its survival that we should take a global view of the problems confronting us. This involves a quest for universal values, which brings us into conflict with religion. Historically religions have evolved over the centuries in different cultures as tribal and regional institutions, with different beliefs, legends, myths, arts and customs; they give cohesion to the communities they serve and provide a way of life within which people can give each other moral support; and this cultural diversity is a source of interest and enjoyment to all humankind. Religions teach us not only what to believe but how we should live; they both postulate beliefs about unknowable mysteries in a dimension outside space and time, and they produce moral codes which they claim are divinely inspired. At the same time, although their teachings are incompatible, most faiths claim that their teachings apply universally and are uniquely true.

We should distinguish between the metaphysical, mythological or divine elements in religious belief and ethical values. The hypotheses of science and the value judgements which guide our conduct in this world can be subjected to the test of empirical evidence and rational debate. But there can be no such rational debate about the metaphysical beliefs of religious faiths. When their beliefs differ and are incompatible, what criteria can be applied to decide which is 'truer' or 'better' than another?

The conflict between the traditional religious faiths and the quest for universal values is a central problem which this inquiry tries to address. For religious beliefs with their unique claims to truth are divisive, and religious extremism gives rise to conflict and can be a source of strife, hatred and violence. There are, however, basic human values which can provide the foundation for a universal creed, a global ethic which can be a unifying force instead of a divisive one. There are many other sources of strife, hatred and violence apart from religion, but in our plural world acceptance of a global ethic can enable, and indeed encourage, the diverse cultures to co-exist in tolerance and peace.

There remains the difficult problem of creating the moral and political climate in which shared values can prevail over divisive religious faiths. The solution must lie in education in its broadest sense, forming moral opinion in the community as a whole, and educating the young to recognise not only the point of view of their cultural group but a wider point of view - the legitimate interest of all humanity.

I am grateful to Mr Brian Pilkington and the Pilkington Foundation for sponsoring publication of this book, and to Dr Terumichi Kawai for introducing it to readers in Japan. Both are dedicated to increasing international understanding between East and West, particularly through inter-cultural exchange in art and philosophical ideas, and I hope that this book can make a contribution to this important process.

I am indebted to Professor Duncan Derrett DD, Emeritus Professor of Oriental Laws in the School of Oriental and African Studies, University of London, for his perceptive and stimulating comments; and my thanks are due to my wife for her patience in deciphering and typing the manuscript. Without her practical help it would not have got off the ground.

Blockley
February 1994

PART I

The Evolving World

Religious Implications of Roy Peacock's book; 'A Brief History of Eternity'

In defining the aim of his book Roy Peacock writes:

> How the universe got there and how it works make for fascinating study. Maybe close investigation reveals a set of fingerprints that show it has been carefully handled but no more. Yet the giants of science thinking in a vacuum bounded only by the experience of their forebears had something else to say. They were interested in how the cosmos works and hence how it originated, but they, like us, also wanted to know why. Many of them found out, and that is the eventual focus. As Hawking puts it [in *A Brief History of Time*] this is knowing the mind of God – and that is the ultimate goal in all science.

His excellent survey of the history of the physical sciences and his explanation of the present state of research is, to me, clearer than Hawking's. One must agree with his general argument that: 'Scientific theories no matter how realistic they may be in our perception cannot be proof of God's existence. We cannot use something of a natural, transient nature to support someone of spiritual eternal nature.' The physics of the cosmos, and indeed all scientific disciplines, do not tell us the why of existence or bring us directly into the supernatural dimension. The world as revealed by science appears to be entirely mechanical, but its ultimate and final causes lie beyond the physical realm of science.

It can be said that in its search for truth science is seeking to reveal and understand the 'things of God'. The aim of Christianity can be said to be the same. Peacock asks: Why therefore is there a gulf fixed between the two? The answer surely is that there is no incompatibility between science and religion. Conflict arises only when on grounds of religious supernatural belief men refuse to accept as true facts and propositions incontrovertibly established by science; or when scientists on the basis of their empirical study of scientific phenomena claim to refute beliefs arrived at not by sensory observation but by a faith which lies within the supernatural or spiritual realm.

Peacock rightly asserts that when Hawking says, 'If we should discover a complete theory [of cosmology] we shall be able to take part in the discussion of the question of why it is that we and the universe exist', he is guilty of a non sequitur. There is no automatic route from the how to the why. Educational as it is to investigate how things happen, why they do is in another dimension.

After his exposition of the 'how' of the physical universe Roy Peacock goes on to explain the 'why'. He argues that there was a boundary condition at the singularity called creation. Before that there was neither space nor time. He concludes from the Second Law of Thermodynamics that there will be in the normal course of events a moment when entropy will have reached a maximum and the concept of time will cease to apply. This is eternity and is the destiny of the natural world. He points out that many of the greatest physical scientists and mathematicians were deeply religious. Pioneering science was a magnet drawing them to God; many, like Copernicus and Galileo, were committed Christians and believed that they had direct knowledge of the 'mind of God', and they had no doubt that beyond the physical world they were uncovering there was a spiritual dimension.

Like Hawking, Peacock concludes that the cosmos is so carefully fashioned that it would be very difficult to explain why it should have begun 'except as the act of a God who intended to create beings like us'. Like Pascal, Peacock underwent a deeply religious experience as a result of which he found knowledge of the God he recognized in Jesus Christ, and like Pascal found this knowledge through the gospel of the Bible. This relationship with God, through Jesus Christ, provides him with an assurance and certainty in the conduct of his daily life, which exists in the area of knowledge which cannot be reached by scientific study. It lies beyond the space/time boundary when the physical laws of the universe will cease to apply, and will continue to exist in the lifeless eternity of maximum entropy. This experience and that of the many great scientists who were drawn to God gives him grounds for believing that this is the key to why the universe exists; it is inconceivable to him that God would have made a cosmos for man to reside in if it were merely to end in maximum entropy, an eternity which will be a 'petrified state with no life or means of living'. He believes that God would intervene to introduce a different eternity. Although we have no scientific means of defining the creation moment or of predicting the end, our recognition of a transcendent God takes us beyond the 'how' to the 'why'.

Roy Peacock's conclusion is that turning to biblical Christianity will suffice to explain the why. Unless the old meanings of the testaments are reinterpreted and remade to be relevant to this modern age I do not think that his conclusion provides a comprehensive 'lesson for our time'. Since the Enlightenment in the seventeenth and eighteenth centuries the tremendous growth of scientific knowledge, combined with Darwinism and the spread of empirical philosophy, has transformed the western world into a secular society, and I shall refer later in more detail to the failure of Christianity, along with the other monotheistic religions, to meet this challenge. The success of science and technology has so dominated the scene that the traditional wisdom and spiritual values of the past have been submerged. Christianity has lost the freshness and vitality which enabled the early church to expand in the remarkable way it did. In order to meet the need of this changing world it will have to renew and reinterpret the message of the gospel in order to make it relevant to the material scientism of our time. Obviously Roy Peacock's experience of God through the Bible is of the greatest importance to him, but the Bible as it is interpreted at present has not proved to be the answer for a great number of people of this generation.

Apart from the fact that Christianity as at present interpreted seems to have lost its appeal and predominance as a living faith having relevance in this scientific age, it must be recognized that religious experience and revelation is a very personal thing and belief is arrived at in a very personal way. Because faith is a psychological fact internal to the believer its validity cannot be refuted. But what happens to one person is not necessarily of general application, and to be universally convincing personal revelations about God must be intelligible and communicable. Many such experiences have been recorded throughout history from many different parts of the world, and it is not unreasonable to explore their sources and the circumstances which gave rise to them in order to test how widely they apply to human experience as a whole.

At this stage it is sufficient to accept Peacock's basic point: that in order to explore why man and the cosmos exist we must break away from the limits of sensory observation and look for any answer there may be in another dimension. There is no doubt that there is a need to pursue this exploration, because without it human life would be meaningless. There does seem to be innate in human beings a need for religious or metaphysical belief in a level of being which transcends the individual and his material existence. People in all branches of science and art, in all walks of life and in all parts of the world, have striven to find something to believe in which is bigger than themselves. The needs of individuals vary greatly and are determined largely by their social, cultural and ethnic backgrounds, but as one writer recently put it: 'Anyone who rejects religion is left with a God-shaped hole.' My purpose will be to see how this vacuum might be filled in a way which is meaningful in our scientific age.

 Revelation and the Mysterious

Before pursuing this inquiry it is worth referring briefly to supernatural revelation in connection with religious belief. There is a large body of literature about divination, prophecies and inspired utterances of sages and mystics throughout the ages. It also covers a very wide range of studies into the supernatural, including shamanism and many forms of psychic phenomena. Study of the structure and layers of the human psyche within the last century or so has also raised questions about the working of the mind which make it irrational to accept uncritically the assertions of those who claim to have communion with God entitling them to speak with divine authority. Such critical questioning must apply to the fundamentalist attitude in various religions, which accepts as literally and incontrovertibly true every word of the Old and New Testaments or the Koran, in spite of the fact that the Christian church claims to be provided by God with a new covenant which supersedes the exclusive covenant claimed by the Jews with their God. Islam in turn claims to be the final revelation of God to man, superseding both Judaism and Christianity. It does not diminish the importance of these religious works; it does however suggest that after centuries in a changing world some reinterpretation or modification is justified in the light of new insights and understanding. In passing it may not be out of place to comment on a psychological aspect of the distrust by some scientists of philosophy and Roy Peacock's references to the links between physical scientists and religious belief. Abraham Maslow, author of *The Psychology of Science*, suggests that the pursuit of science is often a defence. 'It can be primarily a safety philosophy, a security system, a complicated way of avoiding anxiety and upsetting problems. In the extreme instance it can be a way of avoiding life, a kind of self-cloistering.' Similarly perhaps acceptance of an established biblical religion provides an assurance and certainty which precludes agonizing religious reappraisal and doubt..

Religious beliefs and attitudes reflect the times and state of knowledge in the societies which give rise to them. It is obvious that from earliest times different races and cultures have produced varying accounts of how the natural world and men were created. These are based on deep poetic insights; they are usually classed as myths and cannot by their nature be scientific accounts of creation, but their metaphorical or symbolic meaning may give them a significance as important to those concerned as more recent scientific theories of how the world began. What is common to deep personal religious experiences and to theological beliefs throughout history in all cultures is the attempt to explain the deepest mysteries confronting man: where did he come from; what meaning or purpose is there in his existence; what is his place in the natural world and the universe; and what is his future.

These problems have been a central study of science and philosophy, as well as religion, throughout the ages. It is true that confronted with the mysterious or the unknown men have tended to use 'God' as a convenient repository for unsolved problems, until they were rescued by the growth of new knowledge or understanding. But these unknown factors and unsolved problems were of two quite different kinds: some were susceptible of solution through new scientific discoveries, but others which by their nature are beyond the reach of science remain mysteries. The essence of the concept of God must surely be that it describes what is unknowable through normal human faculties. If any knowledge of God is to be obtained it must be through faculties other than the physical senses; and for those who do not have personal revelations the concept does indeed tend to be a repository for the deepest mysteries which can be identified in human experience. As Einstein put it:

> The most beautiful thing we can experience is the mysterious. It is the source of all art and science. He to whom this emotion is a stranger, who can no longer wonder and stand rapt in awe, is as good as dead; his eyes are closed. To know that what is

impenetrable to us really exists, manifesting itself as the highest wisdom and the most radiant beauty which our dull faculties can comprehend only in the most primitive forms – this knowledge, this feeling is at the centre of true religiousness. In this sense, and in this sense only, I belong to the ranks of devoutly religious men.

The Failure of Monotheistic Religions

The explosion of knowledge in recent times through the discoveries in cosmology, quantum physics, medicine and other sciences, is both exciting and of immense-importance to our understanding of the universe, our environment and life within it. But as has been mentioned above, the success of science and technology has so dominated the world in the last two centuries that the wisdom and spiritual values of the past have been submerged. Progress is measured in material terms, and in the endless pursuit of economic growth or personal wealth. The result is that the quality of life and the longer-term benefit of the community are frequently forfeited for short-term gain, and thoughtless over-consumption threatens the future of planet earth. Empirical philosophies have prevailed, and scepticism has tended to undermine all forms of supernatural belief, including religion. Logical positivism is one such philosophy, produced by the scientific attitude which assumes that only knowledge tested, measured and quantified is worthy of attention. It has served a useful purpose in clarifying our use of language and separating the language of science from that of ethics and metaphysics, but apart from this, by asserting (admittedly at its more unsophisticated levels) that only propositions which are empirically verifiable are 'meaningful', it led a whole generation up a sterile cul-de-sac. Marxism has taken over in many parts of the world, postulating that the universe comprises nothing but matter; and its Hegelian philosophy of dialectical materialism, incorporating humanistic idealism, has given it a quasi-religious significance. The reaction has been a rather incoherent upsurge of various movements which are trying to rediscover spiritual values and reunite mankind with the whole natural world. Hence the growth of evangelistic sects, experimentation with psychedelic drugs, the green movement and the spread of 'New Age' groups originating from the western USA.

The monotheistic religions in the western world set up a God who exists outside the material universe, the omnipotent Lord of Creation; they separated Good from Evil and Heaven from Hell and have created an apparently insoluble dilemma of a good all-powerful God being responsible for evil. In doing so they have unwittingly spelt out the basic premise for the dominance of science and materialism. At the same time their promise of a future life in paradise or eternity has failed to alter man's basic predisposition here and now to greed, power, cruelty, violence and short-sightedness. In Britain the emptying churches show that Christianity has failed adequately to relate its image of an ascended and glorified Christ to the times we live in. The main reason for this must surely be that in this changing world people's perceptions of the meaning of God have remained too rigid. The process of historical change slowly empties them of their original vitality and turns them into superstitions. It is futile to lament that old certainties are dissolving away; that is the nature of certainties, and the solution must be to reforge their meanings so that through a process of rebirth and mutations they can provide a living faith. It is not sufficient simply to say that our problems will be solved if we turn to the Bible. Albert Einstein has been quoted above in connection with the mysterious. There is no question that he was a deeply religious man; he considered God to be essentially the sum total of the physical laws which describe the universe, and this view is increasingly that of many people today.

 A Philosophical View of Man and Evolution

The search for a comprehensive philosophy must surely begin with the study of man and his make-up. It is stating the obvious to say that all knowledge and understanding acquired by human beings have come to them through their faculties. Epistemology, how we know what we know, and the significance of human experience have been a study of philosophers and thinkers throughout the ages. What an individual experiences and believes depends on the way his faculties apprehend both himself (his inner world) and the outer world. The way in which we receive and translate experience of the outer world is extremely complicated. Physiologists have shown that sense impressions are themselves constructed by the brain and the nervous system in such a way that they automatically carry with them an interpretation of what is seen or heard. But apart from the use of our senses in acquiring knowledge of the outside world we continually use other faculties, such as insight, illumination and intelligence, without which our observation of the material world would be meaningless.

Physicists use the Anthropic Principle to support the idea that the universe was made for man. In its weaker form it postulates that the physical properties of the universe are so delicately and precisely balanced that if they were different humans could not be here to see it. In its stronger version it assumes that it was made with such precision so that man must evolve. This highly speculative assumption has been countered by Stephen Jay Gould in his book, *Wonderful Life*, in which he demonstrates, on the basis of detailed palaeontological investigation of sites in British Columbia where an amazing diversity of the earliest-known fossil forms exist, that if life were to evolve from its earliest forms again it would probably not do so in the way that it has, and would not end up with the same evolutes. Be that as it may, given the actual instead of a hypothetical state of affairs it seems to me that the obverse of the anthropic principle is no less significant: 'We see the universe the way it is because that is the only way our limited faculties can see it.' If we are to deepen and extend our understanding of the mysteries of the universe, therefore, the starting point must surely be study of man's make-up, what we mean by his psyche, the interrelationship between his 'mind' and physical body, his relationship with the world around him, and how he has reached his present place in the hierarchy of nature. If we are looking for the fingerprints of God they are at least as likely to be found in the biological sciences which study life, nature and humanity in all its diversity, as in the study of the physical cosmos and quantum physics.

In his book *A Guide for the Perplexed*, published shortly after his death in 1977, E.F. Schumacher reviewed the findings of science and religious experience throughout the ages. As a result he has produced a comprehensive system of ideas, examining man's relationship with his environment and also his own evolution and beliefs about the universe. It is particularly relevant to our modern age because it takes full account of scientific knowledge and is consistent with science's endless curiosity and quest for new discoveries. But at the same time it sets out a biological framework within which individuals can identify higher levels of existence and places responsibility on them to develop their talents both to serve the world we live in and to enhance their own spiritual dimension. It leaves men free to adopt such religious faiths as they need to transcend a merely material existence. While it has universal application in the world, it provides a philosophical basis for extending traditional Christian values of justice and peace by respect for what Lord Runcie, the former Archbishop of Canterbury, termed the 'integrity of creation'.

Central to Schumacher's wide-ranging survey is the statement of a Great Truth: that there is a hierarchic structure of the world, which makes it possible to distinguish between higher and lower Levels of Being, and that is one of the indispensable conditions of understanding. Without it, it is not possible to find out where everything has its proper and legitimate place; everything, everywhere, can be understood only when its Level of Being is fully taken into account. The Levels of Being roughly correspond to the four 'Kingdoms' recognized by our ancestors: mineral, plant, animal and human. However the Levels of Being reflect successive gains of qualities or powers as we move from the lower to the higher levels. The difference between inanimate matter (call it m) and a living organism is life. Chemistry and physics have been unable to explain this character or quality (x) which differentiates a live plant from a dead one, and the existence of life represents an ontological discontinuity in the hierarchical progression, or more simply a jump in the Level of Being. From a plant to animal there is a similar jump, a similar addition of powers:

typical, fully developed animals can do things which are totally outside the range of possibilities of the typical, fully developed plant. These powers (y) can be labelled 'consciousness', which can be recognized easily if only because an animal can be knocked unconscious, a condition similar to that of a plant: the processes of life continue though the animal has lost its peculiar powers.

Moving from the animal to the human level one can identify additional powers. What precisely they are is a matter of controversy; but it is an indisputable fact that man is able to do innumerable things that lie outside the range of possibilities of even the most highly developed animals. What does this power (z) comprise? Man can not only think, but he is aware of his thinking. He is capable of watching and studying his own thinking; he can accumulate and store knowledge and let his imagination range into the future and the remotest areas of space. He is able to say 'I' and to direct consciousness in accordance with its own purposes. This mysterious power 'z' opens up enormous possibilities of purposeful learning, investigating, exploring, formulating concepts and ideas, accumulating and communicating knowledge, and of imagination. It can conveniently be labelled 'self-awareness'.

This outline of the Levels of Being simplifies Schumacher's detailed and convincing arguments that there is progression from lower to higher Levels of Being which are ontologically disparate. Each level is obviously a broad band; allowing for higher and lower beings within the band, precise determination of where the lower band ends and the higher begins may sometimes be a matter of difficulty and dispute. However, the essential differences between the levels are differences in kind not degree, between inanimate matter and the powers of life, consciousness and self-awareness; and difficulties of identification and demarcation cannot be used as arguments against the fact that there are ontological gaps and discontinuities separating the four elements m, x, y, z from one another.

Man can thus be written $m+x+y+z$. He has appeared on earth by a process of evolution, and in reaching his present level of self-awareness he has built on the lower levels of existence, containing them all within himself, as physiologists and biologists have observed. Evolution is a term covering all biological change occurring in the constitution of systematic units of animals and plants since the beginning of time. The changes are amply attested by the fossils found in the earth's crust and by the observations and studies of workers who attempt through their various branches of biological science to survey and explain the phenomena exhibited by living matter. As a general description of biological change the process of evolution operating through the mechanism of natural selection for adaptation and survival can be taken as established beyond any doubt, and it puts the changes occurring in plants and animals into historical sequence with a very high degree of scientific certainty. But in the light of present knowledge the emergence of life from matter, and the development of consciousness and self-awareness in living organisms, cannot be explained as the mechanical product of nothing but chance and necessity, i.e. the inexorable laws of physics. Schumacher's detailed exposition of the thesis that the distinctions between levels of being represent progressive differences in kind not degree, and that there are ontological gaps and discontinuities separating them, is of great importance in considering evolution. If evolution cannot *wholly* be explained as the mechanical product of nothing but chance and necessity, then the existence of matter, life, consciousness and self-awareness represents four irreducible mysteries.

In postulating that life, consciousness and self-awareness are ontologically different and that their emergence represents an irreducible mystery, Schumacher specifically rejects the hypothesis of the neo-Darwinists that the world has evolved by a purely materialistic, mechanical process of natural selection and adaptation; and by denying that the process can be adequately explained by scientific investigation he opens the subject to metaphysical speculation. It is a matter of controversy whether the levels of being emerged by a 'seamless' process from atomic particles and molecules or by quantum ontological jumps; at this stage the relevant consideration is that there are major differences between inanimate matter and living, conscious and self-aware organisms. The important fact is that with Homo sapiens the character of evolution has changed. The mechanism of selection, genetic and environmental variation, is no longer exclusively 'natural'; it is distorted by man's ability deliberately to act in ways which affect his own personality and survival, and change his environment and life on our planet. 'Success' in biological evolution can no longer be defined merely in terms of reproductive dominance.

 The Faculty of Self-awareness

We must revise therefore the view of modern materialistic scientism that the cosmos, and hence man, is nothing but a chaos of material particles without purpose or meaning. We do not have to close our minds to 'higher or lower'. If we see man as higher than any arrangement, however complex, of inanimate matter, and higher than the animals, no matter how far advanced, we must see him as head of a hierarchical structure of the natural world, and with an open-ended potential for rising higher. This is the most important insight which follows from contemplation of the four levels of being; at the level of man there is no discernible limit or ceiling. Self-awareness in fact is a power of almost unlimited potential, and opens up the possibility to rise above the levels which at present we regard as human.

We are so familiar with this innate faculty of self-awareness, on which we continuously depend in our everyday lives, that we scarcely appreciate how special a quality it is. Its powers are essentially a limitless potentiality rather than an actuality. There is nothing automatic or mechanical about them. The inequalities in the human endowment cover infinitely varying degrees and kinds of capacity between individuals, and the powers have to be developed and 'realized' by each if he is to become a fully human person. Throughout history individuals of genius have reached pinnacles of creative and spiritual achievement, as inspirers and founders of great religions, and in the various sciences, philosophy, literature, music and the other arts; and these provide standards of attainment to which the rest of humanity looks up. But this creative process is still at work: with his open-ended faculty of self-awareness Homo sapiens has a potential for almost endless development and it is impossible at present to foresee the heights to which he may attain.

 Evolution in Perspective

From these conclusions various important philosophic implications follow; but before going on to refer to these it is necessary to look at the evolutionary process in perspective. We are told that the solar system has existed for some four to five billion years, and that before energy is exhausted in the entropy of eternity, to use Roy Peacock's concept, it has several billion years to go. We are only beginning to unravel the complexities of the human organism, and to acquire knowledge about the world we live in and discover the incredibly complicated interdependence of the whole of nature. There are still immense distances to travel on this journey of discovery. Man has put his fingerprint on the world only for some tens of thousands of years, a brief moment in the time-scale of the universe, and it is surely beyond rational belief to assume that he and the natural world of which he is part have in their progress from lower to higher levels reached a zenith in the evolutionary process. In another few millennia his faculties and knowledge may have transformed his view and understanding of the planet he inhabits, his capacity for managing the natural world, and indeed the very character of that world, if it has not in the meantime been destroyed.

At this point it is relevant to look at the idea that the destiny of the universe is to exhaust itself in the entropy of eternity. To quote Arthur Koestler:

> The gospel of flat-earth science was Clausus' famous second Law of Thermodynamics...Only in recent times did science begin to recover from the hypnotic effect of this nightmare [i.e. that the universe is running down like clockwork, because its energy is being steadily exhausted so that it will finally be exhausted in a state of maximum entropy] and to realize that the Second Law applies only in the special case of so-called closed systems (such as a gas enclosed in a perfectly isolated container). But no such systems exist even in inanimate nature, and whether or not the universe as a whole is a closed system in this sense is anybody's guess.

From the aspect of day-to-day living or making policy decisions affecting the next century or two, it is of little significance what the condition of the universe will be some billions of years hence. But in relation to religious belief it is of some importance to consider what eternity might mean, because we shall see that in the religious sphere the only thing that appears to be common to all religions is that they are in some way connected with 'the eternal', whether this is understood as an eternal Being outside man or as the eternity of man's own being, his immortality. It is reasonable to say therefore that not only are the beginnings of the evolutionary journey hidden behind the big bang, if that is how the universe started (or behind the continuous creation of matter out of nothing, which is the alternative theory held by some scientists such as Fred Hoyle), but the destination is also unknown. When therefore we have referred before to the hierarchy of nature, with man at its head, we should envisage a natural world in which not only are the powers with which man is endowed open-ended, but the future of the whole hierarchy is open-ended.

 Homo sapiens as part of the Evolutionary Hierarchy

What then is it reasonable to deduce from the knowledge we have acquired so far about the natural world and man's place in it? We have seen that at the beginning of time our physical universe was probably formed by an enormous explosion of energy in the form of atomic particles, and after some billions of years by a mysterious process living matter became separate and clearly distinguishable from inanimate matter. Living organisms developed which had the important ability to reproduce themselves and hence to evolve into diverse forms of life and increase so as to cover the earth. Thus a new and different kind of living energy came into being, out of which the natural world evolved over billions of years by a process which can be described as continuous creation.

When consciousness appeared through another mysterious jump in evolution in the form of animal life, this living energy continued to work through the animal kingdom, and following the instinctive mechanism of natural selection an almost infinite number of species developed. But with the appearance of man the course of evolution was transformed; when his intelligence and self-awareness were added to animal instinct he not only incorporated living energy and consciousness within himself but through his additional powers became a self-motivating creative being. His power to decide his own conduct enables him to change and manage his environment; he can enhance or destroy it, and he has therefore a unique responsibility for the condition of the world around him. Having in the process of evolution incorporated all the levels of being within himself he is part of nature, not above and outside it. It is necessary therefore to revise the idea implanted by the Old Testament prophets that nature exists solely for man's benefit, because 'the Lord of Creation put all things under his feet, making him to have dominion over the beasts of the field, the birds of the air and the fish of the sea'. Man must recognize the integrity of creation and the interdependence of all life on earth, and that he should exercise his responsibility accordingly.

We have been looking at 'man' as an abstraction – Homo sapiens – fulfilling a role in the history of evolution. In the real world, of course, there is no such thing; there are only individuals living out their lives, and 'humanity' is only a shorthand for their aggregate. If we accept that humanity represents the highest level so far reached in a progressive evolutionary hierarchy, then each member of the human race has a responsibility to use and extend to the maximum his talents and creativity so as to realize his full potential as a human being, an individual personality. The endowments and gifts of individuals in both kind and degree vary immensely, as do the conditions they are born into and the opportunities open to them. Relatively few can be outstanding and fewer still geniuses, but every human being from the highest to the lowest has a life to live and the sum of their contributions during that life determines how the world as a whole will move from one generation to another, towards higher or lower, forward or in regression.

As has been said above, the potential for man's development is open-ended and seems to be virtually unlimited. It is beyond the scope of this paper to discuss in detail how his powers can be extended but three principles stand out as prime requirements. First, ignorance is the enemy of progress. The pursuit of knowledge is important both for its own sake and because without it enlightened problem-solving is impossible. It is an imperative therefore both to 'know thyself' and to accumulate all the knowledge that can be gleaned of the world and the universe. Secondly, the creativity in man should be nurtured and developed by every means; it is an essential element in all the great discoveries of science, and the driving force of all great literature and art. Without it life is reduced to a static mechanical process of purposeless reproduction. Thirdly, man should seek to expand his mental capacity by every means available to him. There is evidence that he has mental powers which are only dimly apprehended and as yet undeveloped, and parapsychology in all its various forms is at last becoming respectable as a subject for systematic and scientific study. This whole field is wide open; as Sir Charles Sherrington, perhaps the greatest neurologist of the century, said in *Integrative Action of the Nervous System:* 'That our being should consist of two fundamental elements offers, I suppose, no greater inherent improbability than that it should rest on one only. We have to regard the relation of mind to brain as still not merely unsolved but still devoid of a basis of its very beginning.'

This brings us back to the assertion that the quality of self-awareness is open-ended. Many thinkers have expressed the view in different ways that at their present stage of evolutionary psychological development they still have a long way to go in obeying the injunction carved on the temple of Apollo at Delphi, 'Know thyself'. It has been very clearly expressed by P.D. Ouspensky, who states his 'fundamental idea', in *The Psychology of Man's Possible Evolution*, as follows:

> ... that man as we know him is not a completed being; that nature develops him only up to a certain point and then leaves him, either to develop further, by his own efforts and devices, or to live and die such as he was born, or to degenerate and lose capacity for development. Evolution of man...will mean the development of certain inner qualities and features which usually remain undeveloped, and cannot develop by themselves.

Schumacher has expounded on this theme at some length in his *Guide for the Perplexed* and I do not propose to discuss it further here. It is also central to the theme of Teilhard de Chardin's *The Phenomenon of Man*, as we shall see when we consider the subject of religious faith.

 How Free is Man's Will?

We have now come near to the point when we can assess how far the philosophy outlined so far goes towards helping to fill what was called 'a God-shaped hole', but first there are some questions to be answered. When we say that individuals should aim at higher things the implication is that they are free to do so. There is considerable controversy about will and how free it is, and many will argue that our thought and actions are predetermined. To a very large extent this is undoubtedly the case; at a mundane level the choices open to us are obviously limited by the alternatives that are physically possible; decisions are influenced by upbringing, culture and environment, experience and by genetical inheritance. Some neurophysiologists studying the apparent mechanistic functioning of different parts of the brain claim that it exists solely for the processing of sensory information and the production of an appropriate response. This leads them to conclude that a person's will, while it is individual, is not 'free'. In reaching this conclusion Rodney Cotterill, in his book *No Ghost in the Machine*, goes on to say: 'What my senses have observed has been passed through my consciousness, down to subconscious levels of my mind, there to be processed in a manner determined by all that is already established in that inaccessible place, irrespective of whether it got there by genetic endowment or by experience, or by a combination of both.' However as a comprehensive account of how the brain and the mind function this appears to be inadequate, for it does not seem to explain the whole of human behaviour. As Cotterill himself goes on to say: 'I cannot be sure that there are not other things lying in my unconsciousness that are even more significant.'

The individual's faculty of self-awareness, combined with his imagination, enables him to be conscious of himself in relation to a wide range of ideas, hypothetical situations and future possibilities; and when we look inwards into the inner space of our awareness we feel that there is an inner world where thought and imagination are free. This inner space is limited no doubt by each individual's experience and genetic endowment, but this is different from asserting that all thought and creative, inventive ideas are determined solely as a mechanical, causal response to outside influences; and in order to explain the whole range of human activity it seems necessary to assume that there are other internal factors at work. Because individuals are capable of foreknowledge and can anticipate various future possibilities they can deliberately originate movement; in short, they are self-motivating rather than passively reacting. They have powers of reasoning and intuitive insight to evaluate action, and within the freedom of their inner space some people have remarkable powers of creativity and invention, By choosing to reach out to wider fields of knowledge and to communicate with others they are able to extend the area of freedom within their inner space.

As Arthur Koestler has recorded in *The Ghost in the Machine*, a symposium on 'Brain and Conscious Experience' held in 1966 discussed the subjective experience of making a choice that is not enforced, not inevitable, and the question was whether this experience is an illusion. The majority of participants concluded that it is not an illusion; and one of the speakers, Professor MacKay, a computer expert whom one would expect to incline towards a more mechanistic outlook, concluded his paper as follows: 'Our belief that we are normally free in making our choices, so far from being contradictable, has no valid alternative from the standpoint even of the most deterministic pre-Heisenberg physics...'*

Throughout history the relation of mind and the brain has raised philosophical and metaphysical questions, which I do not propose to discuss here. Modern research has demonstrated that there are many mechanisms in the brain which work for the purposes of the mind automatically when called upon. And the question arises, what agency is it that calls upon these mechanisms, choosing one rather than another? Is it another mechanism or is there in the mind something essentially different? Questions such as this give rise to speculation about dualism, the separation of the mind and matter, body and soul. But serious study and systematic exploration of the mind and the brain are still at a very early stage; much remains to be explored, and to make assumptions as to whether the two are one must tend to block the progress of research, not only into the relation of the mind and body but the phenomenon of extra-sensory perception.

---

* Heisenberg's Uncertainty Principle, one of the foundations of modern physics, suggests that on the quantum level strict determinism no longer exists.

To sum up this section, most of us, most of the time, behave and act mechanically. There are many forces of external necessity, accumulated in the past, which determine our actions. But the specifically human power of self-awareness enables us to attain to a level of freedom which makes it possible to establish independent goals and to work step by step to achieve them. This demands concentrated conscious effort, and it applies to us not only as individuals but collectively, because people can combine to form a collective will and achieve collective goals. It will be apparent that this conclusion has important implications in relation to personal and social responsibility, both morally and aesthetically, because it makes it difficult to shuffle off responsibility on to inexorable external necessity.

 Evolutionary Change: Higher or merely Different?

Another question that sceptics may ask is what is higher? There has certainly been change over centuries but is the level we are at now really 'higher' than it has been in the past? Can higher be defined in terms of moral and aesthetic values? Is it more than increasing material prosperity for the community we live in or for the world as a whole? If we think of moral goodness many people today call for a new moral basis of society. But if you turn to ethics as an academic philosophy for guidance as to how to conduct yourself you are likely to get an exercise in verbal gymnastics in reply. For Plato 'the good' was an abstract idea existing eternally apart from the temporal existence but in which human life could partake. For some it is a statistical exercise based on the utilitarianism of 'the greatest happiness of the greatest number'. Again another school of thought gives an account that suggests that all ethical propositions reduce to expressions of simple emotion, whence its popular title of the 'boo—hooray theory of ethics'. In practice, of course, it is impossible to decide what is good or bad, higher or lower, without an idea of purpose; good for what? In dealing with the broad ethical issues and what is .higher or lower in the moral structure or in terms of human values we must have a clear idea of what we believe to be our purpose in life.

What then is the purpose of life on earth? The answer will vary according to the psychological make-up of each individual. Some people find it futile to torment themselves trying to discover the meaning of existence. It is enough to say that life is for living; their personal credo may be to maximise their pleasures and minimise their pain, and bear with stoicism those miseries which cannot be avoided. Such people can lead good lives accepting the traditional values of a Christian society, such as justice, peace, honesty, kindness and charity or love, but at the same time they find that they either cannot or do not need to philosophize, or believe in Christian or any other theology. On the other hand we have followed a path of enquiry where the evidence has shown that Homo sapiens, as a species, is part of an evolutionary process; with his emergence on earth evolution took on a new character. His capability of self-awareness has given him an immense potential for further development; at the same time it has put him in a unique position in the natural world, and his conduct can materially affect its condition.

When we look at what he has achieved we cannot but be appalled: bitter divisions between religious faiths and sects frequently lead to bloodshed; ethnic groups are in savage conflict one with another; nationalisms inspire internecine strife, over which hangs the threat of nuclear war; the growth of industrialization and urbanization may have improved the material well-being of many but it has created immense social problems; and through lack of foresight and proper management of resources the terrestrial and aquatic environment and the earth's atmosphere are being damaged. Confronted with this situation most thinking people will say that there must be some meaning of existence which holds out hope of a better world, however distant the prospect may be. This meaning, this hope can be translated into a belief that there must be more to human life than a more or less frustrated craving for pleasures of various sorts, combined with the experience of pain, ageing and death. This is where religious belief comes in, but before we move on to the subject of religion, let us be clear about one thing: as we live our day-to-day lives we are continually having to solve problems, whatever long-term beliefs we hold about the meaning of existence. Problems involving values are usually divergent, that is they have no one solution, they require a balance to be struck between opposing or conflicting factors, such as justice and mercy, freedom and order, private interest and public interest or social altruism. These problems cannot be solved in the way that mathematical or laboratory scientific problems are solved; they have to be grappled with and their solution needs the exercise of judgement. In short they have to be transcended by wisdom. A set of religious rules or dogmas may relieve individuals at times from the need to work out agonizing decisions, but specific dogmas or rules may not apply in all circumstances, and general principles such as 'love thy God, and love thy neighbour' put the responsibility back on the individual. Dogma can very easily become bigotry.

 The Importance of Knowledge and Love

Earlier we said that the monotheistic religions had failed to adapt to our scientific age; we should now, in the light of the changes that have taken place over recent centuries in the development of ideas and knowledge, consider in what respects religions have become out of touch and how they might be revitalized to meet the spiritual and emotional needs of our times. For the more intellectually inclined these needs may be met by a rational, secular philosophy; but for most people they can be satisfied only by some form of religious belief which gives temporal life on earth a more spiritual and perennial dimension. If religious beliefs are to be credible to the mass of humanity they cannot be at odds with the known facts of the world. To be in tune with modern knowledge and understanding and meet the spiritual needs of mankind any such system of belief should be compatible with the following:

1 The world and universe is in a continual process of evolutionary change. The world is at present divided, geographically, ethnically, culturally, nationally, socially and religiously, and if we are to live and survive together the process of change should be helped and guided by every means to reduce conflict and promote global unity.

2 Humanity and the whole natural world are interdependent. Advance towards global harmony requires respect for living creatures, and co-operation, tolerance and good will.

3 Man is incomplete as a human personality, and as head of the natural hierarchy should develop his powers of self-awareness and creativity for the benefit of the world as a whole.

4 In order to make informed decisions we should aim to increase and expand knowledge in every direction, and as knowledge increases we should be ready to adapt our beliefs to take account of the changing situation.

As Sir Julian Huxley has said in his Introduction to de Chardin's *The Phenomenon of Man*:

> Knowledge is basic. It is knowledge which enables us to understand ourselves, and to exercise some control and guidance. It sets us in a fruitful and significant relation with the enduring processes of the universe. And by revealing the possibilities of fulfilment that are still open, it provides an overriding incentive. We, mankind, contain the possibilities of the earth's immense future, and can realize more of them on condition that we increase our knowledge and our love.

Finally, to conclude this section, in using the word 'love' we should be clear about its significance. It has been so debased in modern usage that it is associated largely with sentimentality, romanticism and sexual appetite. Love is not peculiar to man, and is a general property of all life. In animals it is easily recognized in its different forms such as sexual conjunction, parental instinct and social solidarity, and C.S. Lewis has written eloquently of human love under the headings of affection, friendship, Eros and charity. Its essential element is sympathy, though of course the words are not synonymous. In speaking of ethics Bertrand Russell referred to sympathy thus:

> All the great moralists, from Buddha and the Stoics down to recent times, treated

the good as something to be, if possible, enjoyed by all men equally. Their ethic had always a two-fold source: on the one hand, they valued certain elements in their own lives; on the other, sympathy made them desire for others what they desired themselves. Sympathy is the universalizing force in ethics.

In evolutionary terms we must recognize primarily love's natural dynamism, an energy or power which, in the face of the competition and conflict prevailing in the world, has the potential to achieve the synthesis of individuals and peoples, reconcile elements with the whole, and create unity out of diversity.

PART II

The Religious Dimension

Differing Attitudes to Religious Faith

Earlier we referred to the failure of the monotheistic religions to cope with the growing dominance of science and rationalistic materialism. In Britain particularly the emptying churches indicate that in this changing world people's perception of traditional religious meanings has not adapted adequately to the times we live in, and it was suggested that the solution must be to reforge them so that through a process of rebirth and mutation they can be revitalized. In approaching this we should be careful to avoid being accused merely of heresy. As one writer has argued, the orthodox doctrine of Christianity is a larger and more inclusive thing than heresy. The heretic 'wants to cut up the cloth and make it into nice little suits according to current fashions'. This can of course be true in particular instances, but it would be wrong to use this as an argument that doctrine should not be modified from one age to another in the light of new knowledge and understanding about man, nature, the world and the universe. We must look at doctrine historically.

Christian doctrine became unavoidable when people started disputing what the experience of Christ really meant. The doctrines of the Trinity and Christology and the Nicene Creed were hammered out over a long period some seventeen centuries ago because heresies were being propounded which the early church believed were not true to Christian experience. The same writer said: 'The Church knew she could never capture these mysteries in words but she had to do something.' We should accept the proposition that the interpretation of the Christian experience will change and be adapted to the needs of succeeding ages. Similarly belief and believability are complex things. An American theologian coined the phrase 'The available believable', knowing that what was available to one man's belief was not necessarily available to another. We should operate on the assumption that Christian doctrine is not identical with Christian experience; the latter has an almost infinite capacity for adaptation to the needs of succeeding generations, and it is much bigger than any one generation's statements of doctrine. This evolutionary view of religious belief applies to all religions; they have a long history of growth and development, and any living faith, by the mere fact that it is alive, is a growing thing. When we talk about reforging religious faith we should ensure that we are not just 'chopping up the cloth to make nice little suits according to current fashions', but trying to identify the timeless elements in religion which continue to meet the evolving needs of mankind in a changing world.

We shall see that there are two main streams of religious thought and I shall try to summarize them later. But before doing so I shall refer to a situation described by Don Cupitt in his book *The Sea of Faith: Christianity in Change,* as it provides an insight into two very different types of religious faith. Pascal, who lived from 1623 to 1662, was a gifted mathematician and experimental scientist; as such he was part of the great revolution in man's understanding of the universe associated with Copernicus and Galileo, which reached its first peak with Newton. Descartes (1596–1650) was also part of this revolution, a leading theorist of the new mechanistic science and an uncompromising rationalist. Both men were Christians, but their attitude to their religion was profoundly different. To put it simply Pascal, confronted by the new universe of science, with the infinite greatness of the cosmos above and the infinite littleness disclosed by the microscope, underwent a deep religious experience which convinced him that the answer to the insoluble questions and torments of life was to throw himself on the mercy of God; and that God was not the God of reason, but the God of the Bible in the person of Jesus Christ. As Pascal saw it He alone can satisfy the human heart, and this commitment gave him heartfelt certainty, joy and peace. 'This is faith, God felt by the heart, not by the reason,' is how he put it. Meanwhile Descartes was following another path; he was deeply interested in philosophy and the justification of scientific knowledge, and starting from the famous premiss of his own existence as a thinking subject ('cogito ergo sum') he aimed to establish a metaphysical framework by speculative argument. In his system there were three entities: the thinking and observing mind as a spiritual substance; the mechanistic universe of bodies in motion; and God who guaranteed both the existence of the physical universe and the power of human reason to attain to a comprehensive science of nature.

Descartes was an orthodox Catholic, but Pascal was intensely critical of the way in which he treated his belief in God, as a being for whom he had no further use after he had set the world in motion. Whether this criticism was fully justified or not, it dramatizes the difference between the philosopher's God of reason, who for Pascal was of little religious interest, and the living God, constantly present, who was to be found only through Christ and by the way of the human heart. For Descartes God was undoubtedly an objective entity whose existence he believed was established by a process of reason as a necessary part of his metaphysical scheme of things. But what is the ontological status of Pascal's God? His faith gave him a deep religious seriousness which governed the conduct of his whole life, but his passionate psychological experience was personal to him alone. Whether God exists or not is not the central issue; the certain reality is the psychological state of mind of the believer. As Don Cupitt has pointed out, this contrast is an early example of a puzzle which crops up repeatedly in later years: the claims of theological objectivity and of deep personal religious experience continue to pull in opposite directions in our modern scientific and rationalist age. Roy Peacock's position as set out in his *Brief History of Eternity* is, I think, an example of this dilemma. It also underlies the upsurge of many different evangelical sects emerging in the religious ferment which is evident in our present times.

The Two Main Streams of Living Faith: East and West

Christianity is of course one of many living faiths. I do not propose to try to produce a summary of comparative religions, but in order to get the subject in perspective it will be necessary to identify the broad differences between various religions and for this purpose I have relied heavily on *The Concise Encyclopedia of Religious Faiths* edited by Professor R.C. Zaehner. There is obviously a distinction to be drawn between comprehensive theories about the world, such as have been produced by philosophers over the centuries, and religions. One person can produce a theory of the world, but he cannot invent a new religion; we must therefore deal with the living faiths which are accepted in the world we know and have survived the test of time, and must therefore meet some fundamental need in man. In our context we shall consider mainly religious beliefs, not the ecclesiastical structures associated with them (churches, temples, mosques, their liturgy, priesthoods, organization and the beauty of the music, art and language that support the great religions and are an important part of their practice). As a corollary to this, when we discuss religious belief we must recognize that there are a great many people who may hold beliefs, but who are not publicly committed to being practising members of particular religions or sects.

Professor Zaehner's *Encyclopedia* is divided into two parts, one dealing with religions of what he calls a 'prophetic' type and the other with religions of a 'mystical' or immanentist type. The prophetic tradition originated in the Near East and is represented by Judaism and the two world religions, Christianity and Islam, that derive from it; these broadly comprise the Western tradition. The Eastern tradition is described by Zaehner as the great contribution of Indian thought to the general religious scene; this starts with the primal form of Hinduism, and the fortunes of this 'incredibly complex religious amalgam' are traced down to the present day. He then treats the offshoots of Hinduism, of which the most important is Buddhism in its myriad manifestations. We shall look at these two broad streams of the East and West briefly, because each is dominated by basic but very different ways of thought, and there is a gulf fixed between their conceptions of man and the world in which he lives. A brief survey of this kind must needs be grossly simplified; but it is important to have some authentic understanding of the respects in which the traditions differ, and the generalizations about them are based on the views of acknowledged experts.

The majority of the English-speaking world is either Christian or continues to act on Christian assumptions, using Christian terminology. Religion is thus equated with the worship of God, and the belief in the existence of God as omnipotent and omniscient is taken to be the hallmark of religion. The Eastern mentality as manifested in Eastern religion operates quite differently. Since the rise of Christianity in the West, European civilization has been dominated by beliefs passionately held; toleration and 'reasonableness' have not been in evidence until comparatively recently, and even now, as the history of this century shows, its hold is very precarious. The European mind treats the concept of Truth as of prime importance, the concept that the ultimate realities can be known at least in part and that the possession of this truth is vital; the same applies to those who subscribe to Islam. This is not, however, the Eastern way of seeing things; Hinduism and Buddhism have always tended to regard different religious manifestations as being in their own way aspects of one indivisible truth which cannot be grasped in its essence because it is ineffable. The endless trail of persecution and massacre which has defaced the history of the West must seem therefore not only wicked but incomprehensible.

 The Western 'Prophetic' Religions

As was mentioned briefly earlier, Western 'prophetic' religion originated with Judaism. God appeared to the Jews as an objective reality who chose them out of all the peoples of the earth and entered into a covenant with them. This covenant works itself out in history; God deals with his people, telling them what to do, encouraging or chastising them, but leading them on to their fulfilment at the end of time when there will be a new heaven and new earth. God speaks to his people through men specially selected, the prophets; through them he makes his will known and it is for man to obey and do that will. He must be worshipped as he commands, not otherwise. This is the type of 'prophetic' religion, and the same accents are heard in the Koran as in the Hebrew prophets. Israel alone of all nations claims to be God's chosen people. The Christian Church claims to be the new Israel provided by God with a new covenant which supersedes the old; this however is rejected by the Jews, and since Christianity claims to be the fulfilment of the old Law and of divine origin, a claim categorically rejected by the Jews, we are faced historically with two religions which both claim to be the one true faith revealed by God.

The situation is made no easier by the emergence some six hundred years after Christ of Islam, which claims to be the final revelation of God to man, superseding both Judaism and Christianity. The three religions, however, though disagreeing on what they would themselves consider essentials, are united in one fundamental respect: each claims to be a direct revelation of the One True God to man. Moreover all three, despite their immense differences, agree in this: that God does make himself known by revelation, that his sovereignty is absolute, and that his will must unhesitatingly be obeyed. All three start with the premiss that God is an external, objective reality, the supreme and absolute ruler of the universe, who created it out of nothing. Further they agree that the human person was created one and indivisible, that bodily death does not mean a final separation of body from soul, but that, in order that man's salvation shall be completed, his body must be reunited to his soul at the end of time. It is agreed that man was created for a purpose which will only be fully revealed at the end of time; and death, the result of sin, is not man's final condition but only a temporary separation. Man is composed of soul and body and final beatitude must therefore include the reunion of man's severed parts: an immortal soul is only half a man. The immortality of the soul, though important in all three religions, is not quite central to any of them. Life on earth is regarded by all of them as supremely important; this life is a time of testing, a preparation for the everlasting. It is unique, and can never be repeated; on it depends our eternal destiny of weal or woe. Earthly life, then, is only a preparation, but a preparation of immense importance; it is a deadly serious affair in which all is at stake, for as all three religions agree we will be held to account by the sovereign God for what we have done and left undone. In all this the three religions which go to make up the prophetic tradition agree.

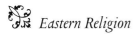
Eastern Religion

What of Eastern religion? After a long evolution from more primitive religious systems Hinduism emerged in the form of mystical doctrines recorded in a series of texts called Upanishads. These varied in many particulars but most tended to emphasize one theme, the unity of the individual soul with one impersonal and absolute World-Soul (Brahman) which pervades and underlies the cosmos. From the time of the earliest Upanishads, perhaps about 600 BC, to the present day this has been the basic doctrine of Hindu mystical religion – the unity of all things in the one Absolute Being and the necessity of realizing this unity within the soul of the individual. Man in his groping towards God is led to experience the immortality of his own soul. At no stage does Indian religion have any clear conception of God as Lord and maker of all things out of nothing, or as an essentially moral being who demands that man should be righteous. The main preoccupation of Indian religion is 'deliverance', that is deliverance of the human soul, which is immortal, from the bondage of the body. Suffering is inherent in human life, which is not regarded as God's greatest gift to man but as a curse which inheres in the nature of things. The Eastern tradition is conditioned throughout by an unquestioned belief in reincarnation, which it regards not as a desirable process but as a supreme evil. Matter, of which the body consists, is a persistent drag on the soul and the soul longs to be separated from it. Thus the essential core of the tradition is to realize a deathless condition in which space and time are transcended and all links with the bodily life have been cut away. The purpose therefore is to escape the cycle of birth, death and rebirth, and to achieve the highest bliss, in which personality is lost in that which both transcends and underlies it.

The Indian religions in practically all their forms would say that this deathless condition of separation of the soul from the body, given the right disposition, can be experienced here and now; mysticism is fundamental to all religious systems and detailed courses of mental and physical training have been devized to induce mystical experience. The experience of 'immortality' therefore is not something we have to take on trust from an external agency.

Professor Zaehner sums up the contrast between the two types of religions thus:

> For the one, man lives but once and it is his bounden duty to do the will of God who reveals himself to him and makes his will known; for the other, man is caught up in an endless cycle of rebirths for which God (if he exists) or Nature is responsible and it is his plain interest as well as his salvation to put an end to the whole ghastly and meaningless round. For the one, God and man confront each other as Lord and servant; for the other, there is no God or if there is one there is no essential difference between him and the human soul.

And again:

> Prophetic religion starts with God and his dealing with man here and now in this world of space and time; the Indian 'mystical' tradition starts with the human soul and the manner in which its release from this world can be achieved. Salvation, whether it is achieved by one's own efforts or through grace of a god, always means deliverance from the bond that links spirit with matter.

It will be seen that the common element in both streams of belief is their preoccupation with immortality: what happens to human life after death? They provide very different answers to this question, but there is another very important difference between them, relating to the way in which they conceive God. Monotheistic, western religions postulate existence of an external supreme being who is both creator and controller of all things, and emphasize the absolute distinction of God from the created order. In the eastern religions, particularly Buddhism, God is nature and nature is God; they interpret all existence as a perpetual and ever-changing flux superimposed upon fixed and immutable laws of nature. In this world spirit is immanent in matter, not separate from it.

The Relevance of Mythology

The definition of mythology given in *Everyman's Encyclopaedia* is 'imaginative traditions concerning the gods and other supernatural beings'. The article goes on to say that the dividing line between myth and legend is hard to draw and behind the purest myth there often lurks prehistoric truth. Indeed, to become established a myth must first be widely accepted as true. The great myths are poetic expressions of early man's profoundest intuitions about the universe and life. A mythology is to some extent a necessity as a background to culture, and even to a reasonably satisfactory human life. Plato while critically demolishing the ancient Hellenic mythology declared that the philosopher would have to invent other truer myths to take its place. Christianity, with Judaism, being an 'historical' religion uniquely offers a myth consisting of historical facts and events; but it too has inspired mythology of the poetic and theological and non-historical, non-factual kind, the stories of Creation, of Eden, of the Serpent of the Trees of Knowledge and of life, of the Rainbow, etc. Mythology then is not (as commonly supposed) confined to an early stage of society, for it still permeates our own, even in the form of non-Christian folklore (e.g. concerning fairies). Another modern example is the Nazi myth; the archetypes of Blood and Soil, of the dragon-slaying Superman, the deities of Valhalla and the satanic powers of the Jews were systematically called up by Hitler to serve the purpose of German nationalism.

The account in *Everyman's Encyclopaedia* goes on to summarize myths under various headings: myth in primitive societies, classification of deities, cosmogony, and Celtic, Egyptian, Hindu, Semitic and Teutonic mythology, and myths of Greece and Rome. It is not necessary for the present purpose to give an account of the myths of different races and cultures. It is sufficient to note that all religions contain elements of mythology. Sigmund Freud wrote: 'The truths contained in religious doctrines are after all so distorted and systematically disguised that the mass of humanity cannot recognize them as truth.' He compares the case to what happens when we tell a child that new-born babies are brought by the stork. Here we are telling the truth in symbolic clothing, for we know what the large bird signifies. But the child does not know it; he hears the distorted part of what we say and feels that he has been deceived. Freud goes on to say that he has become convinced that it is better to avoid such symbolic disguising of the truth in what we tell children and not to withhold from them a knowledge of the true state of affairs commensurate with their intellectual level.

Joseph Campbell probably more than any other individual in recent times has studied and brought together a host of myths and folk tales from every corner of the world; in doing so he has translated their symbolism and demonstrated the parallels that are apparent between them, and has thus developed 'a vast and amazingly constant statement of the basic truths by which man has lived throughout the millenniums of his residence on the planet'. According to Campbell mythology has been interpreted in different ways by the modern intellect: as a primitive, fumbling effort to explain the world of nature (Frazer, an anthropologist); as a repository of allegorical instruction to shape the individual to his group (Durkheim, a sociologist); as a group dream in the collective unconscious, symptomatic of archetypal urges within the depths of the human psyche (Jung, a psychologist); as the traditional vehicle of man's profoundest metaphysical insights (A.K. Coomaraswamy, an Indian philosopher); and as God's revelation to his children (the Church). Mythology is all of these; the various interpretations are determined by the viewpoint of the interpreter. When considered in terms not of what it is but how it functions, it seems to serve mankind in two main ways. First it provides a focal point for social cohesion: ceremonies of birth, initiation, marriage, burial, installation and so forth serve to translate the individual's life-crises or main events into classic impersonal forms relevant to his social group. As all participate in the ceremonial the whole society becomes visible to itself as a living unit. But secondly, as the individual is a member of society, so is the tribe, village or city, or indeed entire humanity, a phase of the immense organism of the cosmos. From earliest times seasonal festivals and magical ceremonies of tribes and peoples have marked the natural changes in the recurring cycle of the year. Rites are also observed in association with great events such as the coming of rain, flood and threats of sickness. In all this the group seeks to make contact with the mysterious powers which appear to control the cycle of the year, the rebirth of life in spring, the harvest, the hardship and periods of joy. Thus the social group and its members try to identify their place in relation to the natural world, the wider horizon of the universe and to the forces at work in it.

As religious beliefs and attitudes have evolved throughout the ages the centre of gravity, that is to say the realm of mystery and danger, has shifted. For the primitive hunting peoples of remote human millennia when the mammoth and the other animals were of primary importance, the source at once of danger and sustenance, the great human problem was the task of sharing the wilderness with them. An unconscious identification took place, which was manifested in tribal totems and in the half-human, half-animal figures of mythological totem ancestors. The animals became tutors of humanity. These myths are still traceable today in the culture of, for example, North American Indians. Similarly, the tribes supporting themselves on plant food became linked to the plant; rituals of planting and reaping were identified with those of human procreation, birth and progress to maturity. Both the animal and plant worlds however were brought under practical and social control. Whereupon the great field of wonder shifted to the skies, the cycles of the planets and calculations of cosmic cycles; man created a sacred moon-king and sacred sun-king and the priestly state which observed symbolic festivals of the heavenly spheres.

As Joseph Campbell says:

> Today all these mysteries have lost their force; their symbols no longer interest our psyche. The notion of a cosmic law, which all existence serves and to which man himself must bend, has long since passed through the preliminary mystical stages

represented in the old astrology, and is now simply accepted in mechanical terms as a matter of course. The descent of the Occidental sciences from the heavens to the earth (from seventeenth-century astronomy to nineteenth-century biology), and their concentration today, at last, on man himself (in twentieth-century anthropology and psychology), mark the path of a prodigious transfer of the focal point of human wonder. Not the animal world, not the plant world, not the miracle of the spheres, but man himself is now the crucial mystery.

The advance of science and secular philosophy, by diminishing and changing the sphere of mythology and mystery and giving rise to religious doubts, has caused profound changes in human understanding and attitudes. One practical result has been to undermine the traditional religions and to lead to an outcrop of forms of belief and worship which are thought by their promoters to be more relevant to the modern age. At the same time the traditional religions seek to adapt to meet this challenge and in the process run into fierce opposition from those who wish to cling to the certainty of the old beliefs and dogmas. In its more extreme instances this opposition takes the form of fundamentalism.

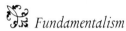
Fundamentalism

Originally fundamentalism was the term applied to the opposition of orthodox churchmen to the teaching of modern science where the latter came into conflict with the Bible story. It came to widespread notice in the USA in 1925 when a young teacher was prosecuted for teaching evolution in a Tennessee state school. Fundamentalists tried to make such teaching illegal in seven or eight other states, but their efforts were killed by ridicule.

The term has now come to apply to radical movements which have appeared in several traditional faiths during this century; these reject modern Western culture and try to restore the old order as recorded in scriptures 3000 to 1500 years ago, both at the personal and the social level. They owe their strength to the fact that they are both pietistic and political, and that they draw on deep emotive forces embedded in their beliefs. They cannot understand the spirit of science and new knowledge, the evolution of a world in process of change, or the modern concern for human rights; but neither can they insulate themselves from them. This tension between religious fundamentalism and the science-based libertarian culture of the modern era is most evident today in the Middle East. The Islamic world in the middle ages embodied one of the world's great civilizations. In recent times for the most part it has been a spectacle of decline and decadence, which has been convulsed by recurrent attempts at cultural revival and episodes of fundamentalist enthusiasm. It appears to be largely rapt in obsolete systems of supernatural and metaphysical faith which are incorporated in the form of Islamic law in the political organization of some states. It recognizes for the most part the secular materialist culture of the modern era as a rival system of belief and as a deadly enemy. Militant political Islam can thus be understood in the more fundamentalist Islamic states as a pathological overreaction to the challenges posed to traditional Islamic culture by scientific, materialist advance in the West.

Moderate Muslims assert that Islam like other religions is adaptable. Only time will show whether the Islamic world can reconcile the sharia with the voice of reason and new knowledge. In the meantime it must be recognized that a new factor has appeared in global strategy. Among the followers of militant Islam the natural human fear of death is subordinated to the prospect of human happiness following martyrdom in a holy war. The logic of nuclear deterrence presupposes that those who possess weapons of mass destruction are dominated by an overriding aversion to self-destruction. But if the leader of a radical Islamic regime, animated by eschatological goals, succeeds in equipping himself with weapons of mass destruction, the logic of deterrence breaks down, because such a leader might be willing to pursue his objectives even at the risk of being annihilated.

When religious fundamentalism is adopted by individuals or relatively small groups of people who wish to pursue a particularly pure, if archaic, version of their faith, they can safely be left to exercise their freedom of choice without doing great harm to society. But where it is imposed on a national scale as part of an authoritarian system, like the mythology of Nazism, the possible implications for evil present a problem of an altogether different dimension.

Religious Pluralism and the Process of Change

Christianity has largely resolved the tensions arising from the conflict between religion and science by accepting that a process of adaptation takes place where metaphysical belief conflicts with new scientific knowledge. T. Corbishley in Zaehner's *Encyclopedia*, while recognizing that at times tension has arisen between 'the Church and Science', has commented that the Church has never been hostile to scientific advance, for it knows that all genuine progress in empirical knowledge is a stage in that discovery of the full truth which is the revelation of God. But there may be a time-lag before the theologian can adjust himself to the new situation. He goes on to say that the kind of problem presented by the biblical account of creation, and the way in which the theory of evolution seemed not merely to contradict the story of Adam and Eve but even to threaten the reality of the divine agency altogether, no longer troubles the theologian. He recognizes the fact that the biblical account was expressing a fundamental truth in a way which would impress it on the minds of the primitive people to whom it was addressed; now he is prepared to see the picture of the origins of the universe produced by the astronomer, the geologist, the physicist and the biologist as much more expressive of the majesty of the Creator than the simple anthropomorphic scenario sketched out in an era of crude thinking. Far from diminishing the value of religion it actually increases the sense of awe and mystery which is at the heart of all worship.

Although this process of adaptation and change may satisfy the Christian theologian it does not seem to satisfy the people of the modern scientific age as a whole. As the area of mystery and mythology has diminished, what remains of the Christian mythos seems no longer capable of making a deep emotional impact on the human psyche. As for the other religions, according to Professor Zaehner everywhere in reawakening Asia the 'intellectuals' are turning away from other-worldly religion, and the success of science in the West and the mastering of the means of production have led increasingly to a reassessment of what constitutes 'reality'.

Hitherto the traditional religions have provided, as well as a religious faith, the social cultures of the societies they dominated. Now not only the metaphysical element of faith is breaking down, but the social framework based on religious dogma is being replaced. One of the clearest examples of this is the changing attitude of both Muslims and Hindus to polygamy and child marriage. Christian dogmatic ethics, as well as beliefs, are also increasingly questioned; whereas in the East polygamy is no longer quite respectable, in the West, particularly the USA, successive polygamy in the shape of easy divorce is becoming increasingly common.

At a recent conference held in London University on the future of theology a lecturer in the history and sociology of religion said that at least 800 'new religions' have emerged in Britain since the second world war, and that if academic forecasts were correct most of Britain's young Christians now in their early teens would have left their churches by the time they were twenty. He said that was possibly because the clergy were presenting the mysteries of faith in a way unacceptable to the young; many former church members had not lost interest but left their church to create their own beliefs. Examples of the new groups included the Children of God, the Worldwide Church of God, the Unification Church, Rastafarianism, and dozens of Buddhist, Islamic, mystic, Japanese, Indian and eastern-style religions. He added that new religions already documented were the tip of the iceberg: 'Below the surface there would appear to be a large mass of new religions which has neither been located nor measured with any precision.'

A major study would be needed to make sense of this outcrop of new religions. It appears that many may not be properly described as religions, but are secular alternatives to religion; controversy surrounds some new groups, involving allegations of brain-washing and mind-control techniques to win recruits; many are dominated by the personality of the founder, leader or guru.

But however ephemeral many of these groups may prove to be they provide evidence that, while there is a shift away from the traditional religions, the need to seek and find a spiritual dimension which can give meaning to life in this material, science-based, humanist age remains as strong as ever.

In this diverse and complex ground swell of interest in religion it is possible to identify some main streams of thought and belief, and at the risk of gross simplification I shall discuss them briefly under the following headings: (1) Convergence of Eastern and Western ideas; (2) Dialectical materialism: Engels and Teilhard de Chardin; (3) The New Age and Green movements; and (4) A religiously plural world.

CONVERGENCE OF EASTERN AND WESTERN IDEAS. With the rapid growth of global communications and the movement of people from one part of the world to another there has been a great intermingling of ideas and cultures. One obvious result has been increased curiosity and information in the West about eastern religions, and this has led to a recognition that the eastern concept of spirituality being immanent in the material world and pervading the whole of nature is very much more in line with western scientific theories of evolution and psychology and with modern philosophical thought than belief in an external Creator and controller of all things, who is absolutely distinct from the created order. We have seen how in terms of both science and mythology the element of mystery has descended from the heavens to man. In the East man and the spirit he shares with nature had been the focal point of religious interest. In the western, monotheistic religions man comprises a dualism of body and soul, reflecting the division of reality into the material and spiritual; and over many centuries language, consciousness and the human mind have been programmed dualistically. This represents a serious difficulty to many people, because without dualism most of what western religion seems to be about would collapse; there would be 'immanentism' without 'the transcendent'. But in eastern theology the spiritual part within man is also part of that which pervades the world; the highest goal for the mystic or holy man is, during his life on earth, to connect his spiritual self with the world spirit and so escape the bondage of the body and the physical world. But for those in the West who have to live practical and mundane lives it is not the ultimate escapism of the mystic which is the primary consideration. Rather, the aim is to enhance spiritual values so as to counter the materialistic attitudes which have increasingly dominated modern life. As this requires development of the spirit within the human psyche and links it to that which pervades the natural world, it provides a bridge in the fracture of reality which lies at the heart of dualism. With the help of techniques such as meditation it offers the ordinary human being the possibility of a new state of consciousness and of fulfilling his full human potential in this life, rather than in an unknown life hereafter. And by linking man to the whole of nature it combines the force of religious belief to that of science and philosophy in providing a vision of unlimited evolutionary and creative potential in the future of mankind.

MATERIAL IDEALISM: ENGELS AND TEILHARD DE CHARDIN. In the early nineteenth century Hegel's ideas dominated philosophy in western Europe. Marx and Engels were amongst his followers and their theories were based on the Hegelian dialectic of opposites, thesis and antithesis, resolving themselves into a new synthesis. Marx's philosophical interests were concentrated on the development of social and economic history and Marxism is commonly understood as the clash between the bourgeoisie, the capital-owning class, and the proletariat, the 'have nots', who have nothing to sell but their labour. Marx argued that this must resolve itself in a classless society in which 'the free development of each is the condition for the free development of all'; the state would die out and an ideal utopia would result. Unfortunately, as recent history has shown, in Soviet Russia the opposite has happened; what should have been a classless society divided itself into a new dialectic of oppressor and oppressed, this time an all-powerful bureaucratic oligarchy on the one hand and the new proletariat, without free democratic rights, on the other. But while Marx concentrated on the secular human problems of State and society and treated religion as of no importance, his collaborator Engels took a much broader view of the human condition and in his later works, particularly in his unfinished work *Dialectics of Nature*, dealt with what are usually considered to be specifically religious issues. This subject is discussed at some length by Professor Zaehner in his *Encyclopedia*, and only a very brief account is given here both to demonstrate that Marxism as propounded by Engels is in fact a quasi-religious faith very akin to the immanentist religions of Buddhism and Hinduism, and also because it has had a marked influence on other current religious attitudes.

In common usage 'materialism' has come to mean an exclusive interest in the creature-comforts of this world to the exclusion of all intellectual, ideological and 'spiritual' pursuits. This is not the sense in which Engels uses the term, and as philosophically Marxism is known as 'dialectical materialism' it is important to understand what Marx and Engels understood by 'materialism' and 'matter'. Professor Zaehner puts it thus:

> Materialism, in the Marxist sense, is simply what philosophers normally call realism, that is to say the acceptance of the phenomenal world as real and the primary datum of all knowledge, as opposed to philosophical idealism according to which the intelligible world of ideas or 'forms' is the model of the 'real' world we experience through the senses. Dialectical materialism, however, goes further than this in that it not only accepts objective reality as given but also maintains, on the basis of evolution, that mind, thought, and all so-called spiritual values derive from matter and are always in the view of science materially based. They are nevertheless the highest evolute of matter and are indeed the goal towards which matter moves.

Matter has in its very nature the potentiality of life, consciousness, thought and ideals. It is not static and mechanical, but dynamic and alive. Human history is simply a continuation on the conscious level of evolution, the course of both being determined by natural law; for matter of its very nature moves from lower forms to higher.

Marxism agrees with Buddhism in accepting that the world we live in – the world of matter – is in a state of permanent flux; everything is constantly changing, coming into being and passing away. But whereas Buddhism declares that man can in this life enter a state of being which 'negates' the world of matter and makes liberation from it the goal of human existence, Marxism accepts the world of matter as the sole existent reality and sees 'salvation' in terms of an eschatological fulfilment which will appear 'in the last days'. The changing world of matter is governed by laws of Nature, by which evolution itself is governed and which are immutable. Although man with the rest of the phenomenal world is subject to the immutable laws of Nature, it does not follow that everything he does, whether individually or collectively, must be fore-ordained. According to Engels freedom does not consist in the dream of independence of natural laws but in the knowledge of these laws, and in the possibility this gives of systematically making them work towards definite ends. Freedom of the will, therefore, means nothing but the capacity to make decisions, or choose courses of action, with real knowledge of the subject.

If by God we understand not the God revealed by the prophetic religions but the unchanging Being against which all change and the permanent state of flux of the world must be seen, then Engels' God is the eternal Law that governs matter, and matter itself, seen as a whole, is the principle of unity remaining eternally the same in all its transformations: 'The form of universality in Nature is law, and no one talks more about the eternal character of the laws of Nature than the natural scientists.' It follows then that the aim of science is to discover as far as is possible these natural laws. Man with his mind or spirit has emerged from matter, and it is the privilege of mind gradually and collectively to come to understand the laws of matter, not as a conqueror but as a partner. Science is a collective affair and becomes more and more so; but however much man progresses he remains an integral part of Nature, his mind or spirit being rooted in matter itself, inseparably interconnected. This leads Engels to say:

> We by no means rule over Nature, like someone standing outside Nature, but with flesh, blood and brain we belong to Nature and exist in its midst, and all our mastery of it consists in the fact that we have the advantage over all other creatures of being able to know and correctly apply its laws ... After the mighty advances of science in the present century, we are more and more placed in a position where we can get to know, and hence to control, even the more remote natural consequences at least of our most ordinary activities. But the more this happens, the more will men once more not only feel, but also know, themselves to be one with Nature, and thus the more impossible will become the senseless

and anti-natural idea of a contradiction between mind and matter, man and Nature, soul and body.

The religious implications of Marxism have never had the same populist appeal as its political and economic theories in Russia and other parts of the world where it has been adopted. It has been unable to develop Engels' pantheistic insights and has remained an intellectual ideology, but not a living faith. It has, however, much in common with Christianity, and Teilhard de Chardin, at the same time a Jesuit father and a distinguished palaeontologist, has reinterpreted it to effect a synthesis between the material evolutionary world and the Christian faith. He entered the Jesuit order at the age of eighteen; but in the course of his education he acquired a profound interest in the theories of evolution. He was appointed to important teaching posts in geology and palaeontology, but some of his ideas were regarded as unorthodox by his religious superiors and he was barred from teaching in France. He spent most of his working life in China, returning to Europe and the USA in 1946; during this time he had written extensively and continued to do so until his death in 1955. In spite of his being a palaeontologist of world renown he was refused permission by the Vatican to publish his work. He was prevailed on to leave his manuscripts to a friend and they could therefore be published after his death, as permission to publish is only required for the work of a living writer.

Teilhard's most important work is considered to be *The Phenomenon of Man*, first published in English in 1959 with an excellent introduction by Sir Julian Huxley. As Teilhard says in his Preface: 'If this book is to be properly understood, it must be read not as a book on metaphysics, still less as a sort of theological essay, but purely and simply as a scientific treatise.' The following very condensed account of this remarkable book does not do justice to the breadth and significance of his thought, but it is included because of the originality of his views and their relevance to the subject of this essay.

Teilhard starts from the position that mankind in its totality is a phenomenon to be described and analysed like any other phenomenon: it and all its manifestations, including human history and human values, are proper objects for scientific study. His second and perhaps most fundamental point is the absolute necessity of adopting an evolutionary point of view. Phenomena are never static; they are always processes or part of processes. Since all evolutionary phenomena are processes they can never be evaluated or even adequately described solely or mainly in terms of their origins; they must be defined by their direction, their inherent possibilities and limitations and such future trends as may be deduced. This applies to the evolutionary phenomenon known as man; as a process in course of evolution he is unfinished and must be completed or surpassed. Teilhard's book is devoted to tracing the stages of man's evolution and deducing the steps leading to his completion.

Teilhard argues in great scientific detail that the 'stuff of the universe', matter, contains within it the germs of everything we understand by 'psyche', and he makes what is clearly for him a crucial distinction in regard to two forms of energy. In addition to energy in the physicists' sense, measurable by physical methods, he deduces from scientific and logical argument that there is also psychic energy.* Whereas the former is destined to exhaust itself, following the principle of entropy, psychic or 'radial' energy, as he calls it, which exists within organized units, increases with the complexity of the units. After the emergence of life in the process of evolution a layer of living matter spread round the world enveloping it so as to create the biosphere, the sphere of life. During the further advance of evolution, awareness, or the mental properties of matter, becomes of increasing importance to living organisms until in man it becomes the most important characteristic of life, giving him his dominant position in the natural world. With the emergence of Homo sapiens the character of evolution has changed; it has become conscious of itself and has assumed the character of a psychosocial process based on the cumulative transmission of thought, knowledge and experience. With the rapid advance in communications and the intermingling of peoples through migration and travel, the realm of thought and knowledge has grown and spread round the world. This has given rise to a layer of mental activity superimposed on the biosphere, and for this sphere of activity Teilhard coined the evocative term noosphere, the sphere of the mind.

In tracing the evolution of humanity from subatomic particles to civilized societies Teilhard stresses the supreme importance of personality. Within the noosphere the individual thinking unit is developing his potential as a human being, and collective humanity continuing the process of psychosocial evolution is steadily converging into an interthinking group based on a single self-developing framework of thought in the noosphere. As complexity within living units and organizations increases, subjective mental energy and activity also increases and intensifies; and extrapolating from the past into the future Teilhard envisaged the process of human convergence as tending to a final state which he called 'point Omega', as opposed to the Alpha of elementary material particles and their energies. Two main factors will co-operate to promote this further complexification of the noosphere. One is the increase of knowledge about the universe at large, from the galaxies and stars to human societies and individuals. The other is the increase of psychosocial pressure from social interaction and

---

* On psychic energy the Revd Dr A.R. Peacocke, author of *Creation and the World of Science*, in a letter to me has commented: 'Teilhard's proposal about two forms of energy has been badly received by scientists, including those (like me) who are deeply concerned about the relation of science and religion. As used in science energy has a well-defined meaning (which it took two centuries of effort to define adequately – and a third (the 20th) to show that energy and matter were interchangeable). Teilhard's proposal is purely speculative and immensely confusing.' Nevertheless the fact remains that the physics which we know at present cannot yet account for consciousness or self-awareness.

convergence on the surface of our planet. The combined result will be the attainment of point Omega at which humanity will have achieved a global unification of awareness, and the noosphere having been intensely unified will have achieved a 'hyperpersonal' organization.

He assumes that this goal may take thousands of years to attain. He does not try to define it precisely, although sometimes he seems to equate it with an emergent Divinity, and at one time uses the term Christogenesis. In leaving it open-ended and vague he is being entirely consistent. As has already been mentioned, he insists that the Phenomenon of Man is strictly a scientific treatise, and in his Preface he also says: 'Beyond these first purely scientific reflections, there is obviously ample room for further–reaching speculations of the philosopher and the theologian.'

Teilhard was a deeply committed Christian; at the same time his experience as a geologist and palaeontologist and his integrity as a scientist made him a convinced believer in evolution. He recognized that there was an affinity between him and dialectical Marxism: in both there is a pantheistic element; both are confident that the future of man lay in his evolution in this world; and they both see evolution as a process leading to an eschatological community which will be free and united in a common aim. For the Marxist this community is the Utopia of the classless society. For Teilhard evolution is the ascent from matter to a community in which there is absolute unity of the spirit. As a Christian, however, Teilhard added a specifically religious element to the spiritual goal which mankind can attain to only in and through matter; he had complete personal faith that his goal could be attained only through God made matter in the Person of Jesus Christ. In this sense he was able to reconcile his radical evolutionism with his Christian faith, and his example offers a bridge between the two. Sir Julian Huxley said in concluding his introduction to *The Phenomenon:* 'We, mankind, contain the possibilities of the earth's immense future, and can realize more and more of them on condition that we increase our knowledge and our love. That, it seems to me, is the distillation of *The Phenomenon of Man.*'

THE NEW AGE AND GREEN MOVEMENTS. The evolutionary theories of dialectical materialism and Teilhard de Chardin have provided an ideological base for various ideas of how 'spirit' or divinity manifests itself in man, the natural world and the universe, and these have greatly influenced the emerging New Age movement. This movement has no priests or institutions, but its followers share a common belief that if humanity does not reform its attitudes we are heading for ecological and spiritual catastrophe. They place responsibility squarely on the individual to take the initiative in this process of reform, but they are steadily becoming more cohesive and organized in the impact they are making on modern society. The initiative in spearheading this movement rests mainly with the better-off for the poorest elements of society inevitably have to concentrate their energies on bare survival.

In Britain in the last decade or two there have been pilgrimages to the ancient sites of Glastonbury and Stonehenge, an annual festival for Mind, Body and Spirit has been held in London, and interest in paranormal and psychic experience has grown. In America sects of the New Age movement attract a large following with their message that divine consciousness runs through everything in creation, and human beings have to clear the channels within themselves whereby they can link the divine element within themselves with this vast interconnecting reality. The goal of New Agers is to awaken and transform their level of consciousness, and various measures of self-help are enlisted to attain this objective, including hatha yoga, meditation, the use of music and 'interiorization', group therapy and holistic medicine. One of the agreed doctrines is 'sufficient is enough': the endless pursuit of economic growth or personal wealth is not only spiritually ignoble but harmful to the planet. When enough people have created their own spiritual reality for themselves, society will change.

The Green movement is an adjunct of New Age thinking. Its members believe that ecological awareness is the outcome of a spiritual journey, and that its demands are quasi-religious, not merely political, in requiring a transformation of human nature. One writer quotes what the Prince of Wales said at the international Ozone Conference in March 1989: 'We thought the world belonged to us. Now we are beginning to realize that we belong to the World.' He commented that this revolutionizes the Judaeo–Christian view of the human status on this planet. It has a religious foundation and represents the irreducible axiom of the conservation and Greenpeace movements. It is not original; it is twenty-four or twenty-five centuries since Gautama (Buddha) taught that this world is 'one seamless garment' and that there is no ultimate dividing line between the man, the tree and the mountain. The British are particularly prone to the idea that God inhabits nature; Blake and Wordsworth expressed the idea in poetry, and there are many instances where spontaneous religious experiences have been triggered by scenes of great natural beauty. Nature can conjure awe where formal worship fails.

The director of a Turning Points forum held at St James Church, Piccadilly, commenting on evidence of a new order being born, said: 'We no longer want to see Jesus Christ held up as the only Son of God, as though divinity was somehow unattainable by everyone else. We want to see him as model of man perfected.' In some of his public utterances when Archbishop of Canterbury, Lord Runcie showed great interest in New Age thinking and expressed the nagging thought that it corresponded better to the spiritual needs of the present generation than orthodox Christianity. His advice to the church was to adopt the New Age agenda, if not the answers; and in a speech to European parliamentarians he suggested the extension of the Christian coupling of 'justice and peace' to a trinity: justice, peace and the 'integrity of creation'. New Agers have no established religion of their own, but they have enthusiastically pointed to the influence of the Bishop of Durham, who in an interview in 1988 described God as 'the power behind things, the presence and the possibility in things and therefore the promise behind things, who is constantly making for love, righteousness and peace'.

Some leaders of Protestant Christianity clearly feel that there is an affinity between the Christian experience and New Age thinking and are feeling for ways in which to adapt and absorb it. But major difficulties remain in the form of doctrinal obstacles, such as how to reconcile an external creator with a form of pantheism, and how to substitute the goal of personal transformation in this world for personal salvation in the next. There is evidence that the churches are facing up to this challenge, but only time will show what the outcome will be.

A RELIGIOUSLY PLURAL WORLD. In a lecture in 1991 the Chief Rabbi designate of Britain, Dr Jonathan Sacks, said: 'I raise the question of how far Christianity and Islam have internalized the implications of a religiously plural world, I wonder whether Christianity has yet come to terms with Judaism as a religion of self-sufficient intent, despite the progress of the last thirty years. What I believe is untenable today is the belief that there is only one path to salvation.' There is no doubt that religious pluralism is here to stay, and in a liberal and democratic country such as Britain it is accepted that religious minorities have a right to their own beliefs; but the Chief Rabbi's questions might more relevantly have been framed differently. If the three prophetic religions Judaism, Christianity and Islam each claim to be based on the truth revealed by God, then their doctrines cannot all be 'true'. And if different religions claim to have exclusive access to divine truth, under what conditions can the great diversity of religious faiths inhabit the globe in peace and harmony?

As has been noted before, religions have in the past served as stabilizing moral centres of the communities they serve, as well as meeting their spiritual needs. With the movement of populations and the growing number of competing religious cults and sects, the established ecclesiastical centres no longer fulfil their traditional function. Different religious dogmas are put forward as representing divinely inspired rules of human conduct, and when they conflict either with the secular laws of the country or with each other it becomes necessary to establish principles which can be accepted as having universal validity and can resolve whatever conflicts may arise. The Archbishop of York, Dr Habgood, in discussing the issue of embryo research put the Protestant Christian point of view thus:

> Why should it be assumed that every moral dilemma has a simple solution? And

why should Christians believe that their faith gives them unique and authoritative insights into problems which are substantially new? A moral response, which allows tentative exploration of new possibilities, with many checks and balances, may be nearer the mind of God, who knows both our strengths and weaknesses, than outright acceptance or rejection. There are of course basic principles on which all people of good will might expect to agree.

The Roman Catholic church with its authoritarian structure and dogmatic utterances cannot be expected to agree with Dr Habgood's liberal view, but if there are 'basic principles on which all people of good will might expect to agree' one hopes that these principles would ultimately prevail, although the process of adaptation and change in dogmatic belief might take longer.

Before considering how far one might establish any such basic principles, there is another factor which should be taken into account. Beliefs can carry strong emotive power, and the creed to which a man is emotionally committed can retain its magic power no matter what evidence is produced to contradict it. Arthur Koestler, in *The Ghost in the Machine,* briefly traces the history of this emotive power from the episode in which Abraham ties his son to a pile of wood and prepares to cut his throat with a knife, then burn him for the love of Jehovah. He questions why so many have for so long approved of this 'insane gesture'. He refers to the religious wars and holocausts perpetrated throughout history as a result of beliefs, and suggests as a possible hypothesis that the delusional streak which runs through our history may be an endemic form of paranoia 'built into the wiring circuits of the human brain'. Be that as it may, one must hope and believe as humanity advances, through the exercise of reason and the expansion of knowledge – especially the perennial therapy of self-knowledge, 'know thyself' – it will come to recognize delusion and paranoia for what it is and will steadily overcome it. As for the basic principles on which all people of good will might expect to agree, it will be profitless to enter into academic discussion about one theory of ethics or another or one set of transcendental beliefs or another, because all have their advocates and they will be prone to be mutually exclusive.

But there are basic principles which can apply in a free and democratic society; they are misleadingly simple and were expressed not by a great religious divine, but by the greatest British philosopher of this century, who also happened to be an atheist. Bertrand Russell when interviewed on television in 1961 was asked: 'Suppose this film were to be looked at by our descendants, like a Dead Sea scroll, in a thousand years time, what do you think would be worth telling that generation about the life you have lived and the lessons you have learnt from it?' This was his reply:

> I should like to say two things, one intellectual and one moral. The intellectual thing I should like to say to them is this. When you are studying any matter or considering any philosophy ask yourself only: What are the facts and what is the truth that the facts bear out? Never let yourself be diverted by either what you may

believe or what you think might have beneficent effects if it were believed. Look only and solely at what are the facts.

The moral thing I should wish to say is very simple. I should say, 'Love is wise, hatred is foolish'. In this world which is getting more and more closely connected we have to learn to tolerate each other, and we have to learn to put up with the fact that some people say things we don't like. We can only live together in that way. If we are to live together and not die together we must learn a kind of charity and a kind of tolerance which is absolutely vital to the continuation of human life on this planet.

However, it is obvious that these basic principles can apply only where there is freedom of thought and a political system to guarantee it. Islam presents particular problems in this connection, and these are referred to later (page 74); but at this stage it is relevant to refer briefly to the war waged in 1991 between a coalition of states, under the aegis of the United Nations, and Iraq, because of the deep and confused emotive forces which this conflict aroused throughout Islamic communities. Saddam Hussein, head of the Republic of Iraq, is one of the least admirable characters in the history of Islam. He was denounced as a 'Zionist agent' by Muslim activists for launching a war on Iran in 1980, which cost an estimated one million Muslim lives; and in March 1988 he used poison gas to kill an estimated 6000 Muslim civilians in one afternoon in the town of Balabja. One might have expected the majority of Muslims and Arabs to have welcomed the attempt by the United Nations to reduce or remove him, as a great service to their brethren in the Persian Gulf. Such a reaction would have shown that Muslims care about the suffering of their co-religionists; it would also have been a step towards friendlier relations between Muslims and the rest of humanity. But this, it seems, was never a possibility. While internecine strife continued between the rival sects within Islam, one factor predominated in this situation: Muslim radicals and Arab nationalists in many countries were united in regarding the Iraqi leader as the champion of Islam, and as a hero because 'he took on the West'.

In the face of such bigotry, fanaticism and hatred it seems futile to appeal to the basic principles of seeking truth based on facts, avoiding hate and accepting the wisdom of love. Nevertheless when we look at the world as a whole and recognize that the globe is pervaded by a universal spirit which is not enclosed by national, regional or cultural boundaries, we can see that there can be no worthwhile future for mankind unless tolerance, truth and love prevail. And we can only hope that in the long term they will.

 The Dilemma of Christianity

This does not pretend to be a learned theological study; it is a layman's view of the situation confronting Christianity in an era of religious pluralism and change. Christianity has of course for centuries been divided between the Roman Catholic, Orthodox and Protestant Churches, and their subdivisions. In recent times there has been a determined ecumenical movement to try to create a unified Christian doctrine, but there seem to be major obstacles to achieving this, not perhaps so much in the interpretation of the biblical testaments as in the human ecclesiology which has formed the institutions and authorities governing the separate churches. In Britain the Protestant Anglican Church is the established religion of the state; its great cathedrals and parish churches exist throughout the land and along with the non-established Protestant churches provide, as it were, the local spiritual and social club for a large section of the population. Even with declining regular congregations these churches provide the centres at which the important events of individual life – birth, marriage and death – are marked by ritual services, and important festivals such as Christmas, Easter and harvest are celebrated, however secularized some of the festivals may be. This great tradition is associated with music and liturgy, art and architecture of great beauty; and even for those who do not regularly attend church services this tradition and the Christian ethic are a spiritual background to their lives.

The Christian tradition in this practical way survives, regardless of internal bickering within the ecclesiastical structure: whether women may become priests, or homosexuality is sinful, to what extent subjective interpretation of the words of the Bible may be permitted, what authority should be imposed on individual conscience in matters of morality, and so forth. But increasingly in its theological beliefs Christianity is becoming alienated from the spiritual life of the community it should be serving. Some leaders of the Anglican Church demonstrate that they are well aware of this. As was mentioned earlier Lord Runcie has expressed doubts as to whether the spiritual upsurge in New Age thinking does not correspond better to the needs of the present generation than orthodox Christianity. The Bishop of Durham earns opprobrium by his liberal interpretation of scripture, which he considers is as true to the Christian experience as more orthodox faith.

The Revd Ian Bradley, author of *God is Green,* argues that three long-held doctrines of the Western church have rightly been seen as major contributions to the environmental crisis which we are now facing: first, the idea that nature exists solely for the benefit of man, which is derived from God's command to man to have dominion over all living things and to fill the earth and subdue it; secondly, the image of a wholly transcendent God separated from and with no continuing interest in his non-human creation, which was encouraged by attempts to distance Christianity from pantheistic non-Christian cults; and thirdly, the darkness which derives from the Greek dualistic distinctions between matter and spirit and more specifically from a particular understanding of the doctrine of the fall. Each of these three teachings, he says, represents a total distortion of the biblical texts from which they are derived and a reversal of the teachings of the early Church Fathers. His views coincide with those of Lord Runcie, who in a sermon at the Festival of Faith and the Environment held in Canterbury Cathedral in November 1989 argued that environmentalism has a spiritual as well as a scientific aspect. As he put it, mankind's pursuit of domination of the planet was an inevitable earlier stage in the relationship between Homo sapiens and Planet Earth; but the time has come to move on to a caring sense of stewardship. Environmentalism, that is to say, has always been implicit in Christian theology, waiting its time to surface.

The impact of recent biological discoveries on religious thought has been very clearly expressed by the Warden of the Society of Ordained Scientists, the Revd Dr Arthur Peacocke, author of *Creation and the World of Science*. In an article in *The Times* on 29 August 1988, he wrote: 'Ever since the discovery of the structure of DNA in 1953 the world of science has been dominated by the explosive growth of the new biology... The discovery of the molecular basis of heredity and of other functions of living organisms, sociobiology, the neurosciences and other developments have generated an entirely new context for our perception of nature, humanity and God.' As a Christian he concluded that 'What the discovery of biological evolution did in Darwin's day, and does ever more so for ours, with the new panorama of cosmic molecular and biological evolution, is to remind us that God is the all-the-time Creator in and through the very processes themselves.' He recognizes that the assumption that 'all is dependent on God who is the creator of time as well as of space, matter and energy' represents a quantum jump from the world of science to that of religion, and requires an act of faith; as he puts it, 'The crunch question becomes: does the "new biology" itself vindicate a materialistic reduction of all talk about living organisms, including ourselves, to talk about atoms and molecules?' Without pursuing his answers to this question, the relevant point to note is that here is an ordained priest convinced that God is a creative force working in and through the whole natural world, not separate from it, and he asks, 'What kind of universe is it if, after aeons of time, the stuff of the world can become self-conscious persons who can think, create and do right and wrong?' This question not only confronts Christians but is central to the religious quest to which humanity has perennially addressed itself.

Perhaps the most outspoken and radical of those who demand a change in orthodox Christian belief is Don Cupitt, Dean of Emmanuel College, Cambridge, and author of *The Sea of Faith*. He argues that religious beliefs, understood as involving supernatural beings, powers and events are manifestly false.

> God [and this is a definition] is the sum of our values, representing to us their ideal unity, their claims upon us and their creative power. Mythologically, he has been portrayed as an objective being, because ancient thought tended to personify val-

ues in the belief that important words must stand for things. Plato, whose thought was half mythological, considered that words like truth, beauty, and justice must designate real beings in a timeless, heavenly world above. His metaphysics was a semi-abstract mythology and we now see that he made a philosophical mistake. Values do not have to be independently and objectively existent beings in order to be able to claim our allegiance.

For Cupitt religion is an activity: it postulates a goal and seeks to attain it. The goal is the ethical and spiritual ideal created by the individual within himself, and for the Christian this means the task of working out a personal vision of Christ, who is our own alter ego, our religious ideal actualized in human form. Religion is thus imperative: a striving to attain to the ideal in this world not the next. It ceases to be indicative – a description of an ideal metaphysical realm which is centred on dogmatic belief, and exists independently of us; such a theology, he says, cuts the nerve of religious striving and asks no more of man than receptivity.

The movement from an understanding of religion centred on dogmatic belief to one centred in spirituality and ethical activity may seem to be irresistible, and Cupitt asks why people then resist it so strongly. He suggests that it is because it is coupled with the admission, at last, that religion is entirely human, made by men for men; this admission is inescapable, and the next stage in the development of religious thought will have to be based on it. There are, however, other forces at work opposing any attempt to adapt Christian belief to the needs of the changing world. Any change is exposed to the charge of heresy. The doctrines of the Trinity and Christology were hammered out over centuries; the process of establishing one accepted doctrine became unavoidable when people started disputing what the experience of Christ really meant, and heresies were being propounded which were not true to Christian experience. The early Church knew she could never capture these mysteries in words but she had to do something. Although this doctrine was established against a very different background of knowledge about evolution, man, the world and the universe, it has acquired the rigidity of long-standing tradition. For many brought up in it, it is regarded as sacrosanct, and like all beautiful traditions it carries its own powerful nostalgia, hence the intense feeling aroused by the liberalism of the Bishop of Durham and other church leaders. It seems to suggest that honest and positive doubt is a sin, and leads to the demand from sections of the congregation for greater authoritarianism and more decisive leadership. Before Dr Runcie made Canterbury Cathedral available for a 'Festival of Faith and the Environment', at which his concluding sermon advocated a change in theological attitudes to the environment and acceptance of the 'integrity of creation', there was an attempt by the increasingly strident fundamentalist element in the Church of England to block the use of the cathedral for this purpose. They felt that the use of church property by members of other faiths compromised the uniqueness of Christianity; but this appeared merely as short-sighted opposition to a constructive move by the church to meet the threat to Christianity from the 'greening' of religion.

In fact Christian doctrine, when freed from its aura of supernatural dogmatism and sterile academic textual studies, has within it a great capacity for adaptation from age to age. The Holy Trinity provides a symbolic, emotive image: in God the Father, an image of what may exist outside space and time symbolizing the answer to the riddle of life in space and time – unknowable but providing comfort to many who feel the need for such an image; in Christ the element of unifying ethical and spiritual values is shown to have entered humanity and to continue to be present as the all-the-time Creator working through the processes of evolution; and the Spirit can be seen as inherent in the stuff of the world, which in the aeons of time since the Big Bang has transformed matter into living organisms and self-aware persons who can think, create and do right and wrong. The principles of resurrection are woven into the very fabric of all life, in birth, growth, harvest, death and rebirth; it should symbolize, not the uniting of the soul and the physical body in a future mythological day of judgement, but the hope for mankind's future in this world and man's place in a holistic natural world. And Christian love, the sympathy of living things for one another, binding them together with tolerance and compassion, is one of the conditions without which there can be no future for our planet.

The Christian churches have a marvellous infrastructure with their long history, cathedrals, forms of worship, priesthood, art, music and so forth; for practising Christians this tradition has great value and should be conserved as the base on which to build and adapt. The dilemma confronting them is whether this ancient infrastructure, with its built-in dogmatism and fear of change because it may be 'heresy', will prove to be such a dead weight in the longer term that it overwhelms the process of adaptation needed to arrest the decline in congregations and meet the needs of the young and of new generations. In Don Cupitt's idea of 'becoming a Christian' the task is like that of the formation of a creative artist. It may take a person years of unremitting labour to find his mature style as an artist; so it should be with the Christian in living his life, striving to realize his own ideal image within his own capabilities and potential. In this he would be helped by the image of Christ, and by the rituals and companions available to him on his journey. But this would mean discarding the option of salvation through acceptance of a religion centred on the supernatural and on dogmatic belief, and substituting for it the hard task for each individual of developing his own spiritual potential and thus building a community of like-minded Christians ruled by ethical and spiritual values. Their goal would be to attain to the 'Kingdom of God' in this life. This interpretation is no doubt deeply shocking to evangelical Christians, but the fact remains that Cupitt's emphasis on the ethical and spiritual elements within humanity is more in tune with philosophical and scientific thought, and with the current upsurge of religious interest than orthodox belief.

It took the early Church some five hundred years to clarify and establish Christian doctrine. Time is needed to allow religious belief to evolve and adapt in a changing world. The question is not a simple issue of defining God in terms of *either* dogmatic, supernatural ontological belief *or* human values. What is required surely is to accept positive doubt as the mechanism whereby Christian doctrine can evolve and adapt so as to remain relevant to the world we live in. Whether, given time, the ecclesiastical structures of Christian churches will permit this process to take place in a way which will command the support of the communities they serve remains to be seen.

PART III

An Holistic Philosophy

The Unifying Potential of Global Values

This essay began by discussing the religious implications of Roy Peacock's *A Brief History of Eternity,* and suggested that, though his experience of God through the Bible is of the greatest importance to him, the Bible as it is interpreted at present has not proved to be the answer for a great many people to the question why the universe exists. This was the starting point for an analytical study trying to trace why the teaching of the Bible no longer commands the almost universal acceptance by western civilization which it has had until fairly recent times. What has emerged from this study? The main fact seems to be that the Old and New Testaments were written at times in history when knowledge of the phenomenal world was at a relatively elementary level. Biblical mythology naturally, therefore, reflects in poetical language doctrine and story-telling which no longer make an emotional impact comparable with that which it made on its contemporary generations. The main difference between then and now is that it was entirely believable that the biblical God should be an external supreme Being not only creating the world and the universe, as it were fully developed, but controlling and ruling it and all its creatures. The development of philosophical thought, growing understanding of the eastern immanentist religions, the formulation of the theory of evolution, and discoveries of astronomy, quantum physics and biology in recent times, have transformed our understanding of the mysteries of the universe. The earlier mythology has thus become increasingly alienated from real life.

The world and the universe are in process of evolutionary change. The whole of nature on planet Earth has evolved from the initial stuff of the world and is continuing to evolve through the mechanism of natural selection and adaptation. In the seamless hierarchy of nature Homo sapiens shares in the element of mind/spirit which pervades it; he is part of the world, the world does not belong to him. With his emergence evolution became conscious of itself, and mankind became responsible for its own destiny and for shaping the future of the planet. But mankind as a species is far from complete; its potential is immense, its future is open-ended and its destiny as yet can only be dimly foreseen.

In this scientific age not only knowledge but the spiritual life of humanity is in the process of evolutionary change. The sphere of mystery and religious belief has moved from the imaginative mythology of many centuries ago to the forces perennially at work in the universe, nature and man. As the boundaries of knowledge are extended, scientists who penetrate the awe-inspiring infinity and beauty of the universe, and discover the wonderful properties of living matter, not surprisingly feel that they are near the 'mind of God', as Stephen Hawking put it. Similarly as we absorb the works of musicians and writers of genius, and look at the way in which great artists give us insight into the world we live in, we feel that we are lifted out of ourselves to a higher level of being than that of our mundane existence. We feel that through the works of human genius we are given a glimpse of the divine. This element of the divine is not the mythological God of the prophetic religions who exists in a heaven outside the world; rather it is the creative spark which is working within men and which is part of the invisible 'spiritual' force which pervades nature.

However, this view of mankind must be tempered by realism. Although mankind has great potential for good, it also has great potential for evil; history is filled with instances of man's inhumanity to man, and destruction both of his fellow men and his environment. In his evolution into the future this conflict between good and evil, beauty and ugliness, is bound to continue. But when we discard the mythology and legends we are forced to recognize that the conflict is not between external divine and satanic forces; it is a struggle within the psyche of man himself, which only each individual can resolve. The psychic energy or spirit pervading the phenomenal world is neutral; man has the potential to channel it towards peace, justice and global unity, or towards divisive strife and a retreat into barbarism. This conflict exists not only at the personal level. As civilization has spread humanity has become divided into innumerable groups and factions – geographical, ethnical, cultural, nationalistic, social and religious. Like individuals these factions are motivated not only by ideals but by aggression, greed, envy, violence, deceit and so forth, with the result that they continually war against each other. If mankind is to live together, rather than die together, the problem is how to overcome the strife which threatens to destroy it.

Religious fervour is one of the deepest and most powerful emotions. It can create both social cohesion and social conflict, and throughout history the great religions, as well as serving as spiritual rallying points for the communities they serve, have been marked by an endless trail of persecution and massacre. The only other comparable motivating forces are, perhaps, loyalty to a leader and national pride and dedication to national interests. But religion, if it takes a world view, has the potential to rise above temporary dictators and nationalism. In particular areas where religion is closely linked to nationalism, for instance when Israel and Zionism are combined with ethnic Judaism, and Arab nations with Islam, the problem of peace becomes so fraught with emotion that it seems almost intractable. Nevertheless, surely the time has come to make a determined effort to seek and find ways in which religious attitudes can be channelled so as to become a global unifying force instead of a divisive one. The issue is whether religion can be revitalized so as to fulfil this role.

Religions are created by humans for humans; they exist to try to satisfy the apparently instinctive longing of humanity to believe that there is a reality which transcends the phenomenal world, and they postulate beliefs about mysteries in a dimension outside space and time. Thus they try to fill the gap represented by the concept of 'eternity'; to this end different cultures have created their own ideal projections of what is ineffable and cannot be described in the language of space and time, and the projections have been given names: Jehovah, Holy Trinity, Shiva, Allah, Brahman, Nirvana, Sunyat and so forth. These beliefs claim to answer the question why the universe exists. But the answer is unknowable by us. The supernatural beliefs of religion are the product of the human mind and imagination; they fill a psychological need but they cannot be held to be 'true' in the sense that propositions about what exists in space and time are true. Many people are absorbed by the question what happens to the individual when life ends. However, no one should think that death is his end; he lives on in the phenomenal world in the influence his life has had on others and in his genes. In the realm of mythology and poetic ideals any dreams are possible: 'Let each man hope and believe what he can.' But let us be clear about this: these dreams and beliefs cannot justify dogmatic assertions of metaphysical truths, nor the strife and warfare that they appear to provoke.

Religions try to answer not only the question what should we believe, but also how we should live. They claim to appeal to what is highest and noblest in the human spirit, and in so far as they speak of the ideal values and behaviour by which we should conduct our lives, such as truth, goodness and justice, tolerance, sympathy and love, such a claim is not unreasonable. All the great moralists, from Buddha and the Stoics to recent times, treated the good as something to be enjoyed by all men equally. But religious institutions and sects have produced moral codes which they claim derive their authority from divine revelation and must be obeyed by their believers accordingly. Because of this they cannot be subjected to the test of new evidence and practical wisdom, and acquire the rigidity of dogma; though, if they cease to command obedience, the institutions themselves are liable to crumble away.

Historically religions have evolved as tribal and regional institutions to serve the needs of different cultures at different periods; but now with the growth of communications and the movement of peoples we can see the world as one interdependent whole, in a way which earlier generations could not. We can see that religious institutions have been divisive and, instead of providing a unifying spiritual force for the benefit of the world, have in fact been a major cause of disunity and strife. Surely the time has come therefore to shift the focal point of religion from earlier myths and outdated supernatural beliefs embedded in the organized religions to what is the nobler side of religious belief, that is the universal ethical values expounded by the great moralists and commonly accepted by all men of good will throughout the world. The spiritual element in man can then provide a unifying force instead of a powerfully divisive one.

These values transcend selfish and factional interests and provide directional beacons: first, for individuals in their personal conduct, secondly for the social groups and cultures that individuals are part of, and thirdly for achieving harmony and cohesion in the planet as a whole. Because language uses words as shorthand symbols for abstract ideas we have been programmed to think of ideas as having some form of independent, objective, timeless existence separate from the phenomenal world. In this we have been mistaken. Words expressing values have no meaning or existence other than as qualities of things, conduct or events in the real world. The relevant question should be not 'what are values', but what conduct do 'value' qualities describe. In short, values can be identified only through experience inherent in the world. They are to be valued for their own sake; they do not depend on the supernatural ideas and beliefs of religious institutions. The sum total of values thus represents the spiritual ideal which should form and direct human conduct, and the aim of humanity should be to ensure that this ideal prevails when it conflicts with evil.

Religion then becomes an activity and a quest. It demands a spiritual transformation of attitude in this world, starting with each individual and extending eventually to humanity as a whole. This represents a superhuman task which transcends the individual and gives life on earth its meaning without recourse to the supernatural. The diversity in the psychological make-up of individuals is immense; it follows that their mental attitudes and energies, and the ways in which they express their religious and spiritual lives, vary immensely. In this diversity the common goal should be to seek and identify the perennial values of goodness, truth and beauty, as guides to behaviour. For the individual this requires demanding work; he is, as Don Cupitt puts it, like an artist finding his own mature style. It takes time and effort; and so in day-to-day living each individual should use his mental and spiritual capacity to identify and increasingly follow the values by which he lives. Because Homo sapiens is a social animal his values must serve to fulfil not only his potential as a self-aware, creative individual, but as a member of various social groups, as a citizen of planet Earth, and as belonging to the natural world. In this quest the essential requirements are knowledge and truth, and tolerance or love, without which it is impossible to reconcile elements with the whole and create unity out of diversity.

Looking at the evolving world we can recognize the interdependence of the whole of nature, including Homo sapiens. We are told that the universe has billions of years still to run; we cannot know the future of the natural world and mankind, but their potential is open-ended, and the possibilities are immense. At a scientific level the world proceeds according to apparently immutable physical laws governing the universe and the biological mechanism of natural selection, and these evolutionary processes are aesthetically and morally neutral. But with the emergence of self-awareness in Homo sapiens the situation changed, and we can identify other factors at work: matter is now defined by the physical scientists in terms of energy, and we can recognize that in this energy there is a creative or spiritual element, which pervades the world and has given rise to persons capable of achieving marvels of scientific discovery, creating artistic works of genius, and telling good from bad and right from wrong. It is this faculty of self-awareness which enables man to identify values and develop his creativity and knowledge. Mankind will have to revise its understanding of value concepts and how they apply to human conduct from age to age, but for our time the values commonly accepted by all men of good will throughout the world appear to provide the meaning of life on earth, and hope and faith in the world's future.

 Global Values and the Established Religions

The idea of spiritual energy pervading the world has much to do with the immanentist Eastern religions. These reject the dualism of body and soul, material and spiritual, creator and creature, which is an integral part of the prophetic religions; in Eastern theology the spiritual part of man is also part of the spirit which pervades the world. These religions are steadily gaining ground in the West because, although their goal, as we saw earlier, is to escape from the material world and merge with the world spirit, many people see the possibility of redirecting or adapting their mysticism and spirituality so as to support movements which regard the natural world and mankind as one whole, and whose aim is to achieve a moral order on this earth, evolving within man and embracing the whole world.

The prophetic religions on the other hand are based on divinely inspired revelations. Their great contribution to the life of mankind has been to produce moral codes to enable communities to live together; these codes require total obedience from their members whose conduct is subject to the oversight and ultimate judgement of their eternal God. But these codes were produced many centuries ago and were related to the knowledge and conditions of life of those times; it was not unreasonable then to believe that man was commanded by his God to dominate the earth and that the natural world should exist for his benefit. Much of the ethical doctrine and creeds of these religions remains of universal application and Judaeo–Christian morality forms broadly the basis of ethical conduct in the West. However, these doctrines and creeds differ as between the different religions; parts conflict with each other and also with new secular ethical judgements based on new scientific knowledge and the realities of life in our changing world. It is worth perhaps digressing briefly to consider whether, and if so how, religious creeds are being reconciled with new knowledge and the values of our modern era.

Earlier it was suggested that Christianity, based on the teaching of Jesus

Christ rather than the institutional doctrine and practice that has grown up around it, has great capacity for adaptation from age to age (see page 63 above). However, the Christian community is not a homogeneous entity; its history is marked by schisms which resulted, firstly in the split between the Roman Catholic Church and Eastern Christianity centred in Constantinople, which in time gave rise to the Eastern Orthodox Church, and secondly in the Reformation in the early sixteenth century which gave rise to Protestantism in direct antagonism to the Church of Rome. The capacity for change and adaptation varies greatly as between these different communities. The Roman and Orthodox Churches are in different ways authoritarian. In discerning between truth and error the Roman Church Christians possess an organ of infallibility in the person of the Bishop of Rome, the Pope. Orthodox Christians believe that the Holy Spirit guides and protects the Church from all error; every Christian has to search his own heart for the right answers, but in order to distinguish between the voice of his God and his own opinions or imagination a Christian has to take account of the opinions of other members of the Church. In practice this means that only those decisions and opinions which have been accepted unanimously by the whole body of the Orthodox Church have divine sanction for Eastern Christians.

The Roman Catholic Church over 1500 years has built up a powerful organization centred on the Vatican and the Bishop of Rome but with distinct local hierarchies. In 1962 the Second Vatican Council was concerned to prevent the Church from fossilizing into a self-perpetuating and inward-looking bureaucracy, and to set it free from traditionalist attitudes and complacent stagnation. According to T. Corbishley's account in Zaehner's *Encyclopedia,* the Council showed a refreshing openness of mind and an appreciation of the many values which existed without benefit of clergy. One of its features was an unwonted reluctance to condemn out of hand any view regarded by the Council as unacceptable. It showed a readiness to appreciate the historical circumstances in which liberal and progressive ideas develop and, when 'errors' arise, to distinguish between the error and the person who errs, 'since the latter is always and above all a human being and must always be treated in accordance with that human dignity'. It was, at least in principle, the death of the Inquisition and its whole spirit.

Soon afterwards, serious tension between Papal authority and opinion within the Church manifested itself in the sphere of the Church's doctrine concerning sex and marriage. A Papal Commission was set up to look at traditional teaching in relation to the population explosion, the problem of world hunger and changing social attitudes in the more highly developed countries. The Pope issued an Encyclical, which ignored the Commission's findings and said, in effect, that he saw no reason for modifying in any significant way the Church's condemnation of any form of artificial contraception. The reaction to this was that, while a large number of clergy and laity welcomed the statement as confirmation of the authority on which they had come to rely, a considerable proportion found the encyclical unconvincing in its arguments and regrettably cavalier in its treatment of the Commission's findings and the views of important Catholic theologians. In country after country meetings of local hierarchies, while expressing their loyalty to the authority of the Pope, yet found it necessary to try to relate the encyclical's teaching to the realities of modern life; they were also unhappy that the contribution of the faithful at large to forming 'the mind of the Church', which had been acknowledged by the Second Vatican Council, had been completely ignored. As Corbishley puts it: 'What in fact was happening was not so much a crisis as a revolution. Old attitudes of unquestioning acceptance of official pronouncements were being replaced by a more responsible and critical estimate of all the factors involved. The primacy of conscience which had always been accepted, in principle, was now being recognized and put into effect in practice.'

Protestantism in rejecting the authoritarianism of the Roman Church placed great importance on the freedom of the individual to interpret the scriptures and to rely on the inner strength and unifying power of the gospels. Since the Reformation Protestantism has coincided with the secularization of European society, and in most countries with the almost total emancipation of economics and politics from religious control. During this period various Protestant communities and sects, too numerous to mention here, have come into being, but it is possible to identify two opposing trends: the liberalism inherent in Protestant faith emphasizes reliance on the sincerity and faith of individual Christians, while on the other hand there is a desire for more centralized authority in the interest of greater cohesion and to provide a link with the great and ancient traditions of the Catholic Church.

Today the liberal tradition of Protestantism seems to be gaining the upper hand. It accepts our modern secular culture as a fact of life and believes that Christianity must learn to dissolve itself into this culture. This means making Christian values the guiding principle within secular morality; it entails emphasizing regeneration of the individual rather than strengthening traditional forms of piety with a strong role for priests and a powerful church organization. The future of Protestantism is not clear. There is some recognition that environmentalism has a spiritual aspect. A gospel and culture movement seeks to rebuild Western intellectual heritage on Christian foundations, and Dr John Habgood, Archbishop of York, in a June 1992 lecture, 'Finding a moral heart in Europe', suggested that such a movement might emerge within the edifice of the Protestant Church, whose aim would be to seize the moral high ground by identifying and reinforcing Christian values in modern society.

Islam is based on the revelations uttered by the Prophet Mohammed in Arabia in the seventh century and collected shortly after his death in the volume called the Koran. The impact he made on the life and thought of Muslims is such that he has been regarded by them as a more than human figure. The Islamic religion provides rules for the whole of life. In its long and complicated history there has been a close affinity between religious and political government, with tension between obedience to civil law and promotion of personal and collective obedience to the Law of Islam, the Sharia. This tension continues; only in the Turkish Republic has the jurisdiction of the Sharia been completely abolished in favour of legal civil codes; in other countries the Sharia structure is still in use to a greater or lesser extent, with orthodox codes being modified in various ways by new laws.

Professor H.A.R. Gibbs described, as one main factor distinguishing modern Islam from its classical civilization and culture, the invasion of foreign influence and ideas represented initially by Westerners employed in Muslim countries in administration, commerce and education; this has led to the adoption of Western systems in all these fields by the Muslim rulers and governing classes, and to the rise of new professional classes trained in Western schools. The reaction to this has been the rise of new associations which have as their main object the defence and consolidation of Islam against the invasion of foreign influence and ideas, and the result has been to make more acute the clash between the secular power and religious leaders, who do not accept that secular law can override the Sharia. He goes on to say:

> Any innovation, propounded by a particular group and made binding by the secular power, is a usurpation of the spiritual rights of Muslims, and carries with it the danger of disrupting the Sharia and consequently of destroying the peculiar and divinely ordained constitution of the Muslim Community. The question for the future is whether Islam will remain, what it has been in the past, a comprehensive culture based on a religion, or become a 'church', a religious institution accepted by larger or smaller bodies of adherents within the framework of a secular civilization.

Political events in the past two or three decades have had the effect of increasing extremism and fundamentalism in some Islamic states, where there seems at present to be a fanatical over-reaction to challenges posed to traditional Islamic culture by scientific, materialist advance in the West. In general the prospect of Islam adapting itself to embrace world values in place of the Sharia in the foreseeable future seems to be remote.

The problem of change of belief within Judaism was very clearly stated by a progressive and liberal Rabbi, Sidney Brichto, in an interview published in the *Financial Times* on 23 November 1991. Speaking about the loss of faith in the Jewish community, he said that the fundamental issue that divides Orthodox and non-Orthodox Jews is 'whether you believe Jewish law is given and immutable or whether law develops over the centuries. Progressives believe in continuous revelation. We can't believe that there has been no progress since the Law was given to Moses.' And about Britain's irreligion: 'Society has inherited the best part of the Judaeo–Christian tradition. Secularism's values in some sense are far superior to the values of traditional religions, because they have made democracy and free expression of the spirit their basis. Religion still has an authoritarianism and puts loyalty before freedom.' On the question 'What is a Jew?' he said: 'A Jew is someone who accepts his Jewish identity for historical reasons and wishes to maintain it...Jewishness is not racial (there are Chinese and black Jews), it is more of a large family.' The Rabbi went on to say that there is a revolution in Jewish psychology with the creation of the Jewish state of Israel. Suddenly they were no longer the perennial victims; that state is the beginning of the fulfilment of Messianic dreams, albeit created by secular forces. But the antithesis of Orthodox and non-Orthodox continues at the wider metaphysical level, because the Orthodox will not accept an Israel restored to them by a human agency.

In this brief review of the main religions in relation to morality, several factors emerge which have important implications for establishing global values which can be universally accepted. First it raises the question whether, to put it crudely, when important moral issues arise such as embryo research, abortion, attitudes to love-making, the use of artificial contraception, and family planning, God or man should decide the issue. Or to put the question slightly differently, should the answer be given by one religious sect or another in the light of its traditional beliefs and doctrine, or should society arrive at a view generally accepted on the basis of the conscience of people of good will, combined with rational free debate, taking account of all the factors involved. Secondly, the moral systems of each religion, like their metaphysical beliefs, differ from each other and in many respects are mutually exclusive. At the same time each, claiming to be uniquely true, competes evangelistically with the others; the result is that they are divisive and often the result is inter-religious strife and persecution. In some countries there is a close link between religion and state, and local religious hierarchies have shown as passionate a nationalism as their political counterparts; too often religious fervour and fundamentalism are exploited to buttress nationalism and increase conflict between communities or states. Thirdly, there is an important respect in which religions are failing the world. We can see now that man's pursuit of domination of the planet for his benefit has been a major contribution to the environmental crisis we are now facing. Although religious leaders pay lip service from time to time to respect for all God's creatures, their moral systems do not aim to redress the damage that has been done; they fail to embrace all life on the planet and our environment with its diverse resources. Their focus is still on the future of man's soul in an unknowable eternity, rather than the future of our living planet.

A new global ethic is needed, which will build on the best that we have inherited from the past, but will extend beyond religious groups with their divisive faiths; such an ethic would view the natural world as one interdependent whole, and its aim would be concentrated on enabling it, with mankind as its self-aware head, to realize its potential not only in the short term but throughout all future generations.

A World Ethic

At a secular level the world's consciousness and energy have been expanding to embrace the whole globe in a way that individual religions have not. Religious eschatological goals were determined by the limited knowledge of the world and the universe available at the time; man's imagination therefore reached into mysterious eternal dimensions – ideal heavenly space and a timeless era in which mortality has been overcome and life is everlasting. Now we can see that, with the revolution in the sciences of quantum physics, cosmology, biology and evolution, and psychology, the old eschatological goals no longer satisfy human intelligence and imagination. While religious metaphysical ideas remain frozen in the past, the frontiers of science now impinge directly on the traditional mysteries: for example, in the field of quantum physics some scientists are prompted to compare the energy which comprises the universe with the world spirit of eastern religions, others to postulate that the laws of physics and the finely balanced chemistry essential to life on this planet require the hypothesis, in some sense, of an intelligent universe; the United States National Aeronautical and Space Administration has embarked on a programme to search for intelligent life elsewhere in the galaxy; with the discovery of DNA and the molecular basis of heredity and other functions of living organisms scientists working in the life sciences have found a new context for the perception of nature; modern psychology is in its infancy, the relation of mind to brain is unknown and we have no clear idea as yet of the possible limits of man's mental experience. Some scientists, such as Carl Sagan, speculate as to how advanced computerized intelligence can be used to make full and creative use of our human intelligence; we are a scientific civilization, they say, and knowledge is our destiny.

International understanding is rapidly growing through travel, media communication and intercultural exchange; our world is now one of international business, electronic communications, global science and technology, and tragic imbalances in the availability of wealth, food and clean water. One of the great unifying forces is the growth and interchange of scientific knowledge. Academic and other intellectual organizations are not confined within narrow national and linguistic boundaries; scientists in every field confer at international level, their findings are published and there is cross-fertilization of research throughout the world. Many non-political international bodies exist to promote better understanding about particular subjects, and intercultural exchange is steadily increasing. At intergovernmental level the United Nations Organization exists for the purposes of maintaining peace on the basis of justice and of promoting friendly relations and co-operation in all matters. The UN are also committed to 'the collective encouragement of respect for human rights and fundamental freedoms for all, without distinction as to race, sex, language or religion'. It provides a forum for handling a wide range of global problems through its specialized agencies and by sponsoring conferences and conventions on important matters such as protection of the world environment.

As man's activities spread across the globe we can see more clearly the interdependence of its diverse parts and the growing understanding between different cultures. We can accept that knowledge and creativity are crucial to the future of the world, but they are rudderless without a common spiritual dimension; it is not scientism and industrialism in themselves that are the cause of the world's troubles, it is lack of spiritual values which can guide knowledge and behaviour toward a better future for the world. Alongside the pursuit of knowledge therefore our scientific civilization should set another goal – a world ethic and global moral values, based on the spirit which is immanent throughout the whole of nature including man.

It is the quest for knowledge about the universe and man's place in it that commands the interest of the present generation rather than the unknowable field of ontological speculation. How this quest can be combined with developing the potential of mankind for its ultimate good is, as we have seen, the subject of religious and philosophical theorizing: for example, a Marxian materialistic Utopia (now discredited); Teilhard's idea of spiritual convergence in the noosphere on Point Omega; or the New Age movement which, if one discounts its more disreputable, anarchic and fringe elements, seems at its serious core to be in close sympathy with the spirituality of Eastern religions, by linking human consciousness with an immanent world spirit. In the light of all that has been discussed so far we can now see the outline of an holistic philosophy, which can provide an ethical framework within which individuals may develop their personal potential, and peoples and nations will have as their common goal the future well-being of the world. Before this philosophy is discussed in more detail it can be summarized thus:

The earth is one interdependent whole. The factor which gives the global ethical system its coherence is the recognition that the natural world, including man, is bound together by the energy which is the stuff of the world, and from which life, consciousness and human self-awareness have emerged. It is akin to the world soul or spirit which is common to the great religions.

Personal and social relations within mankind. Building on the moral values which we have inherited from the great moralists of the past, the global ethic will transcend the moral codes of exclusive religious communities. It will be based on moral principles which can in time be accepted by all the world's diverse cultures, enabling them to live in peace and harmony.

Mankind's relationship with the rest of the natural world and the environment. The religious tradition that man is entitled to dominate the planet for his benefit will be superseded by his awareness that he is part of the biosphere which is planet Earth, and that this requires due respect for all life, and use of the planet's resources in such a way as to be available equitably for all peoples of the earth and to be sustainable for all generations.

The ultimate goal is to realize mankind's potential and the possibilities of the Earth's immense future. This goal will replace mythological religious beliefs. The final end of man is as yet unknown, but his future on earth will be based on his ability to identify the values of goodness and beauty, and to establish them as beacons to guide the way into the future. This will mean developing the spiritual element in his make-up, which is the source of creativity, knowledge, wisdom and love, for it is these which give value to life on earth. We are confronted with an open-ended, mysterious and exciting future which stretches into several millennia ahead, and although we cannot see the final end it is enough that we have a spiritual dimension to point us in the right direction.

If this world ethic were accepted generally, what difference would it make? Human nature is what it is – greed, selfishness, lust and aggression will continue to be opposed by kindness, unselfishness, compassion and reason. But its acceptance would mean that attitudes would change. In the West we have been programmed to believe that goodness derives from God; but when we say that God is goodness, God is truth, God is love, we are saying in effect that love, truth and goodness are crucial to life on earth. The concept of an eternal God in heaven is embedded in the mythology which we have inherited from the past; it is so deeply ingrained that to reject it is likely to inspire in many people feelings of guilt or fear of divine wrath. Nevertheless the time has surely come to recognize that it is up to mankind itself to deal with its problems, instead of appealing to an external deity. The transformation of our view of divinity will revolutionize our idea of moral responsibility because it puts the responsibility fairly and squarely on human beings, individually, in groups and in the mass. We can no longer attribute success or great calamities to an external deity who can confer joy and happiness but also terrible suffering. We have to accept the hard work of dealing with these problems, whether personal, social or national, ourselves. However we have immense resources to draw on. Firstly, we can recognize that spiritual energy is immanent in the world; it is akin to the world spirit accepted in different forms as part of the metaphysical beliefs of the great religions, as World Soul, Brahman, Logos – call it what you will. Individual human beings share this spiritual energy with the rest of nature, and it is this that enables them to be creative and inventive, to know good from evil, and strive after beauty rather than the squalid and ugly. Some time in the future quantum physics may explain it in terms of particles and mathematical formulae, but it will remain nevertheless an hypothesis required to make life on earth worth living, because it will not explain why we can look at nature and great art, and see it as beautiful and moving rather than a mass of subatomic particles dancing in a chaotic flux. In marshalling this spiritual energy for the good of the world, people will continue to use the techniques which have always been used by humanity to gather their spiritual strength and summon up their inner resources: prayer, meditation, yoga, and devices such as music and art, which reach the mind but do not require words to do so.

What does this boil down to in practice? For most people life is about personal and social relationships; without love and sympathy life is scarcely worth living. But problems concerning these relationships are usually divergent, that is to say the factors involved pull in different or opposing directions and they do not have single, simple solutions (see page 34 above). When there are conflicting sides to a problem the solution will depend on trying to reconcile them, using knowledge of all the relevant facts combined with reason, wisdom and sympathy. But moral problems have a long history, and when we consider the possibility of establishing universally acceptable moral standards we should build on the foundations which we have inherited from the great moralists of the past. In doing so we should distinguish between moral rules or codes and the moral values which underlie them. With the increase in knowledge and education thinking people are no longer prepared to accept without question the moral rules and directions of religious institutions. As circumstances and people change in the evolving world old problems change and new problems arise; inevitably they will be subject to free debate and reappraisal, and old perennial values will be applied to new situations. It is this process of reappraisal which is the basis of a global ethic; it will not be tied to exclusive revelations made to religious communities many centuries ago, instead it will take into account all the factors involved and arrive at solutions through the exercise of reason, wisdom, good will and conscience. This process of reappraisal and free expression of the spirit can provide a moral consensus for our multi-faith societies. Such consensual judgements may correspond with religious doctrine, but when they do not religious institutional authority will wither away, and be superseded by a morality which can be accepted by all cultures.

This is not the place to open the Pandora's Box of questions about how we should judge 'diseases' such as drug addiction, alcoholism, anorexia and various pathological conditions. Suffice it to say that we should beware of substituting scientism for morality. Psychologists, doctors and sociologists may describe the conditions in which these diseases occur but, without trying to make any judgement about individual cases, it is surely not unreasonable to make the general point that morality requires hard work, self-discipline, unselfishness and consideration for others. These virtues need not conflict with realizing each individual's potential; indeed they are an essential part of his or her humanity – as a member of the human race. In this scientific and industrial age it is more important than ever that the spiritual element reflected in moral values be built into education and mass-media communications.

For leaders of social groups and countries the concept of personal responsibility has great importance, for as their sphere of influence is greater than the individual citizen's, so is their responsibility. Powerful political leaders often claim that they are under divine guidance. Let us explode this myth once and for all; they can draw on spiritual energy and put it to good use or bad, but they are subject to the same self-discipline, social discipline and morality as the rest of humanity. But in the last resort their power depends on the individuals who belong to the communities under their control. As the old aphorism has it: 'Every country has the government it deserves.'

Global Values and the Environment

Turning to the relationship of humanity with the environment and the rest of the natural world, we are confronted by problems of ethics and global values which have arisen in their current acute form within the past two or three decades. The facts are fairly clear: they were presented by the United Nations World Commission on Environment and Development when in April 1987 it gave a solemn warning that human activities were for the first time surpassing the Earth's capacity to cope with them. In order to understand the issues involved and their moral implications, it is necessary to refer briefly to the Commission's report, *Our Common Future*, and its conclusions. The report says that the exploding world population, more than five billion and due to double by 2050, and the world's industrial output, which had increased fourfold since 1945 and would probably again increase fourfold or more, were causing global environmental damage that could not be sustained; both were putting intolerable strains on planetary systems, in the atmosphere, in soil, in water, among plants and animals, and in the relationships of all these. Global warming, ozone depletion, desertification, large-scale pollution and loss of species were all threatening to combine with extreme poverty and hunger in the third world to create a crisis which could destroy 'the security, well-being and very survival of the planet'. Its conclusion was that the world was at a turning point.

Two main trends emerged in the report: the new environmental problems that were emerging were global rather than regional; and much of the development process in the third world, with its current aid programmes, was clearly failing. The Commission concluded that environmental concerns and poverty were linked and should be seen as two sides of the same problem: in Africa, Asia and Latin America farmlands were being destroyed by destitute farmers who overexploited them out of desperation; population expansion created pressure on land use and would increase the threat of global warming, but desperately poor people need the security of large families. In this analysis the Commission also concluded that it was possible to do something constructive and that was, instead of turning off all economic growth, to aim to achieve sustainable development, which it defined as development that meets the needs of the present without compromizing the ability of future generations to meet theirs.

The issues raised in *Our Common Future* have been brought into the open at the UN Conference on the Environment held in Rio de Janeiro in June 1992, the professed purpose of which was to get the world to change to a path of sustainable development. This is an extraordinarily high ambition, because it means asking people to change the way they do things everywhere. It implies new economic policies for states and a change in habits for individuals; and as a general economic policy it certainly means abandoning the pursuit of simple economic self-interest. But according to the report there is no alternative for the world.

The publicity given to this World Summit has highlighted the various conflicts of interest which have been thrown up by its agenda. They are made immensely complex by the sheer diversity of the countries of the world, varying from primitive tribes in the forests of South America, through developing countries of the third world, to wealthy industrialized countries at the forefront of technology and international trade. It is not necessary to enlarge at length on these conflicts of interest here; suffice it to say that green policies cost money: felling some of a forest and replanting is more expensive than clear-felling; foregoing the use of destructive pesticides means a sacrifice in crop-yields; halting the growth of the motor vehicle traffic that increasingly contributes to destruction of the atmosphere means stringent fiscal or other controls, though everyone wants to own a car. The result has been to set rich and poor nations squabbling; the rich have done the damage and if the poor are to bear the brunt of being environmentally friendly they expect to be compensated. While the rich countries are sitting complacently on their wealth and prosperity, the poor countries still have a bargaining counter, because they own large areas of the world's environment. The World Summit has served a crucial function in focusing the world's attention on the global problems of mankind and the environment and has produced a fair consensus on the kind of political policies required to deal with them, but in practical terms, after all the rhetoric, the question still remains whether the political will has been generated which is needed to make bold practical decisions concerning the world's future.

The World Summit must be seen as merely the beginning, though a very important one, of a process which one hopes will gradually produce incremental gains. There will be all kinds of obstruction because the process certainly means abandoning the pursuit of simple economic self-interest; industry will use its power to prevaricate, the blinkered will try to discredit green activists by labelling them green freaks, eco-terrorists and so on, and although lack of family planning has obviously not been a cause of the population explosion, the Roman Catholic Church continues to proscribe, for doctrinal reasons, the means of contraception which could help to moderate it.

In order to generate the political will required for bold decisions what is required is a transformation of attitude, whereby the more prosperous individuals and nations accept an element of sacrifice of their standards, so that in time the planet's resources can be shared more equitably with other peoples of the world and by future generations. As the report *Our Common Future* has put it, our materialistic consumer-oriented society has indeed reached a turning point, and unless we change direction there can be no long-term future for the world. The change in direction involves non-materialistic, 'spiritual' values which not only make people's experience richer but enable them to take an altruistic view of the world as a whole. As a first step, instead of unrestrained and irresponsible pursuit of material advancement the guiding principle should be 'sufficient is enough'. The issue now is the survival of the planet, and values can no longer be reduced to purely quantitative economic concepts. If, as *Our Common Future* says, human activities are for the first time surpassing the Earth's capacity to cope with them, and what people are now doing in the world is unsustainable and will lead to global disaster, then economic arguments about extra cost, and questions of 'marginal cost', 'exchange value', 'use value' and so forth, cease to be of primary importance. We must add to economic advantage another dimension, which involves altruism, transcends narrow personal and national interests, and makes the future well-being of the whole world its goal.

Because of the urgency of the environmental crisis it receives wide and continual publicity, and the movements which make environmental matters their concern have a high profile. The way forward in forming public opinion is undoubtedly through well-informed and active organizations, such as Greenpeace, the World Wide Fund for Nature, Friends of the Earth and other conservation bodies, who lead the way in performing this function. The United Nations has embarked on an Environment Programme (UNEP) which includes a world-wide programme of education about the critical position in which the planet finds itself. Education in all its forms and mass media communications are clearly the most important means of reaching the world's population, and gradually they and the environmental movements associated with them will influence public opinion.

However, in the other area which has been identified in the world ethic outlined on pages 77–8 above, concerning personal and social relations, the issues are far less clear cut and do not lend themselves to dramatic public debate. Moral problems have changed with the times but they have been with us for thousands of years. In recent times they have been seen as the gradual suppression of religion and spirituality by the juggernaut of material advancement, in our high-technology industrial society. But we can now see the two sides of the world ethic have a common centre: the future of mankind and the natural world depends both on shaking off the mythological supernatural eschatology of the old religions and on rejecting the unrestrained consumerism of our materialistic society, and replacing them with a new faith in the mind/spirit which is immanent in man and the whole natural world, and on which hope for the future well-being of the world rests. This involves a transformation in attitude and a change in direction; it is bound to take time and to achieve it is indeed a tall order. But it is the way forward towards a harmonious and civilized future.

An Holistic Philosophy in Perspective

Many people brought up in the monotheistic tradition who have deep religious beliefs will question whether an ethic based on individual responsibility and conscience can serve as a moral foundation for mankind; they may in good faith believe that without an overseeing God, with the power to mete out rewards or penalties on a final day of judgement, mankind will disintegrate into moral anarchy. Again, people talk with nostalgia about the loss of spirituality in modern society; it is not clear exactly what they mean, but broadly it seems to be regret that people no longer hold the old metaphysical religious beliefs and that standards of conduct are increasingly based on the pursuit of material wealth and status. Throughout this study a central theme has been that we can find the spiritual dimension in life not outside the phenomenal world but in the mind/spirit which is immanent in it. Man is able to recognize its existence through his faculty of self-awareness, which also enables him to identify values whether moral or aesthetic and distinguish them from the neutral atomic structure of the world of physics and chemistry. Consider music as a simple illustration of what is meant; it can be described by the physicist in terms of energy, subatomic particles and wavelengths, arranged in particular patterns in space and time, but when it is listened to it can be transformed into sounds of great beauty, and the listener can not only experience the sounds as such but is aware of himself doing so. This mental experience is 'value-added' to the physicist's analysis, and will vary greatly according to the listener's capacity to appreciate music. So it is with the whole range of man's experience;

it is his ability to add value to the atomic flux which is the world of the physical sciences that gives meaning to life. The great religions have aimed at supplying man with a sense of worth or value, not only in his capacity as an individual but as a member of a community, to which he is expected to contribute at least as much as he takes for himself; with the growth of scientism and industrialism this sense has been largely submerged by the rising tide of materialism, with its emphasis on quantity and material status, rather than on quality and spiritual worth. The problem which now confronts us is to restore the balance, not merely in particular communities but in the world-wide context of global values and the environmental crisis, discussed above.

The world's population is now more than five billion. In this bewildering diversity the mental and spiritual capacity of individuals varies immensely, and relatively few are actively interested in changing either their own beliefs and attitudes or those of their fellows. The motivating force for the great majority is looking after their own well-being and that of their families and friends, and in many poverty-stricken parts of the world life is, tragically, a desperate and often losing struggle for bare existence. For most people therefore life is for living from day to day within the constraints of their circumstances and the framework of law and order in the societies of which they are part; their 'spiritual' aspirations are limited, in various personal ways, to being decent citizens and enjoying leisure occupations, gardens and hobbies, unambitious activities which are none the less laudable for that.

At the other end of the spectrum of spiritual experience is that of the mystics. Those of us who have no direct experience of this form of introvertive exploration of man's inner life can only rely on others who have studied the subject. According to Professor W.T. Stace, in his book *Mysticism and Philosophy*, the basic psychological facts about the mystic's experience are in essence the same all over the world in all cultures, religions, places and ages. They consist in excluding from consciousness all physical sensations, then all extraneous thoughts and sensuous images, then all abstract reasoning, thoughts and volitions. One supposes that then all that remains is emptiness, a void; but mystics, thousands of them all over the world, unanimously assert that on the contrary 'what emerges is a state of pure consciousness – pure in the sense that it is not the consciousness of any empirical content. It has no content except itself.' This experience is akin to a central teaching of all the great religions, which in their various languages urge man to open himself to the 'power' within him, to transcend empirical consciousness by self-awareness. This whole area of experience remains shrouded in mystery for the great majority who are uninitiated, and little serious attention has been given to the methods of studying the inner world of the mind. Mystics warn against being misled by 'corruptions', brilliant lights, feelings of rapture and other sensory experiences, which are obstructions in the path to pure consciousness. We are told not to confuse occult phenomena with true self-knowledge, but the basic question, how the stream of consciousness of the empirical ego interacts with the pure consciousness which appears to underlie it, is unexplained. As yet there is no agreement by neurologists and others concerned as to whether it will be possible to explain the mind on the basis of neuronal action within the brain or whether our being is to be explained on the basis of two fundamental elements. Regrettably, at our present level of understanding, this field of self-awareness seems to be wide open to speculative musings and abstractions, leading to the formation of movements and sects whose leaders seem in some cases intent on various kinds of emotional and commercial exploitation.

At this point it will be useful to turn to the world of physics, as an example of the place of science in the life of the world. We can dismiss as irrelevant Stephen Hawking's view, quoted at the beginning of this work, that if we could discover a complete theory of cosmology we could find the answer to the question of why we and the universe exist. However much we observe and explain how phenomena work, to assume that such knowledge will explain why they exist is a delusion. However there are some interesting and important inferences to be drawn from the spectacular advances made in modern physics. Fritjof Capra, in his book *The Tao of Physics*, has explored the parallels between modern physics and Eastern mysticism. The physicist begins his enquiry into the essential nature of things by studying the material world. Penetrating into ever-deeper realms of matter he has become aware of the essential unity of all things and events. 'In modern physics the universe is experienced as a dynamic, inseparable whole which always includes the observer in an essential way. In this experience the traditional concepts of space and time, of isolated objects, and of cause and effect lose their meaning. More than that, he has also learnt that he himself and his consciousness are an integral part of this unity.' While physics approaches reality from outside through the phenomenal world, the mystic conducts his exploration through the inner world of consciousness at its various levels and discovers a reality behind the phenomena of everyday experience; like the physicist he is aware of the whole of the entire cosmos which is experienced as an extension of the body. In his exploration of the parallels between modern physics and Eastern mysticism Capra concludes:

> The modern physicist experiences the world through an extreme specialization of the rational mind; the mystic through an extreme specialization of the intuitive mind ... Both of them are necessary, supplementing one another for a fuller understanding of the world ... Mystical experience is necessary to understand the deepest nature of things, and science is essential for modern life. What we need, therefore, is not a synthesis but a dynamic interplay between mystical intuition and scientific analysis.

In our society this interplay has not been achieved, and in the last few centuries scientific, materialistic attitudes have increasingly predominated.

Western society has favoured self-assertion over integration, rational knowledge over intuitive wisdom, science over religion, competition over co-operation, expansion over conservation, and so on. It is clear however that a reaction is taking place, and movements are evolving which bear witness to the change. In response to the threat to the environment we have seen that many organizations are concerned with ecology and conservation. There is increasing interest in alternative medicine and holistic approaches to health. There is much talk of the New Consciousness, round which movements are emerging. In this connection the label New Ageism is regarded with suspicion in England, because of the way it is often reported in the media, but its central theme is a belief in a consciousness which brings all people together; New Agers hold that in the course of evolution man removed himself from experiencing directly the energy which pervades the world, but we are now on the point of connecting with it again. This does not entail rejecting the benefits of technology; it means restoring the balance between materialism and higher values.

In this perspective global values and an holistic philosophy have a great part to play. By providing positive aims which can be accepted as being of overriding importance for the world as a whole they can provide a foundation for peaceful coexistence by nations, and they can be a unifying factor at the centre of the immense diversity of cultures and ethnic divisions. However, to expect people to change their beliefs, attitudes and habits, which have been ingrained over generations, is a tall order. It will take time and effort. Education will have a vital part to play, extending beyond culture-centred learning to include the wider resources of the human cultural heritage; mass communication in its various forms and free democratic discussion will also be crucial in forming world opinion. Art can also make an important contribution, because great art helps to develop our higher faculties, and it will combine with great literature and with folklore to enable the holistic philosophy to reach the collective unconscious of mankind. Many will dismiss efforts to establish an holistic philosophy based on global values as naive, impractical idealism. Idealism it certainly is, as in a sense are all long-term strategic objectives; it may take many generations to realize, but one thing is certain: unless we identify common objectives to guide our path into the future and work to achieve them, the prospect for the world, if it continues on its present materialistic and divisive course, is bleak indeed. Let Shakespeare, as so often, have the last word: 'The end crowns all, and time that old common arbitrator will one day end it.'

 Conclusion and Summary

Whether our holistic philosophy might be called a 'religion' will depend not so much on abstruse theological arguments as on how we use words. If religion is defined as faith in what ultimately concerns us in this world, to provide individuals with a meaning in life which transcends the self, then faith in global spiritual values and creativity, as providing a guiding light for humanity in its evolution towards future goals as yet not clearly known, can be regarded as the basis of a religion. But if religion is to be defined as faith in an unknowable supernatural 'reality', then our faith in what ultimately concerns us in this world would be called a philosophy for living.

When people are asked what is the purpose of life, many will reply that it is the pursuit of happiness. But happiness is not something that can be pursued as an end in itself; as Aristotle pointed out many centuries ago, happiness is a by-product of 'the good life'. It is like the bloom on a peach; if there is no fruit, there is no bloom. Again, when we talk intellectually about making the pursuit of knowledge, creativity and global values the basis of human life in place of supernatural eschatologies, we should avoid the mistake of throwing the baby out with the bath water. The mythology of mankind, including the myths on which the great religions are based, provides deep insights into the human psyche. Much of it is recorded in great literature; the story-telling, music and art of religions are a precious part of the human heritage, and their beauty and ability to enrich human experience should not be discarded. It is not the myths themselves, but the fanaticism and fundamentalism that are associated with religions and nationalistic myths that have potential for evil. Similarly the concept of a universal mind/spirit throughout the world, and a common ethical goal of global harmony, peace and justice should not mislead us into assuming it to be the same as uniformity of culture or centralized political authority. Cultural diversity is a precious source of interest and aesthetic enjoyment. As we learn about the arts, literature, architecture and so forth of other cultures we are enriched and made wiser; and sport between countries both serves as an outlet for national competitive fervour and at the same time brings their peoples together in common pursuits.

When all is said about religion and philosophy, we shall continue to be surrounded by the miraculous. The whole universe and our tiny planet represent an infinite miracle – a source of endless wonder. The awe-inspiring depths of the cosmos, and the complexity, savagery and beauty of nature are continuously being transmitted to increasing numbers of people through modern mass communications and education. With the advance of knowledge and practical experience from earliest times the realm of mystery has shifted stage by stage over the millennia from the animal and plant worlds and the great field of wonder in the heavens until the last century, since when the focal point of wonder has concentrated on man himself; and man is now the crucial mystery. Who knows what man will be exploring in future millennia? But at this crucial stage of the world's evolution there can be no more supremely important subject for exploration than man's mental and spiritual capacity and his interdependence and interaction with the living and spiritual energy which pervades the world.

Finally we can condense what has been said before into a few short paragraphs:

1. In the past religious belief has centred on unknowable supernatural entities or deities, such as an all-powerful, omniscient creator of the universe who intervenes in the affairs of men. In the last few centuries our knowledge of the world and the universe has been revolutionized by scientific discovery – in the fields of cosmology, quantum physics, biology and evolution, genetics, psychology and so forth. As a result, for increasing numbers of people religious interest now centres on discovering how the potential of mankind and the living world can be fulfilled on earth. The goal is seen to be salvation or redemption in this world rather than another unknowable dimension.

2. In advancing his own evolution man should use the faculties which distinguish him from the rest of the animal kingdom: first, the ability to use creative energy to expand knowledge and invention, and to create and appreciate works of literature, poetry, art and music of great beauty; and secondly, his ability to distinguish good from evil and right from wrong, and to direct creative energy into channels which will benefit, and not harm, the world.

3. Matter, the initial stuff of the universe, is defined in terms of energy, and contains within it the seeds of life and creativity. In the course of evolution it has given rise to man along with the rest of the world, and the creative spark in man's psyche has made him a self-motivating person capable of using this energy for particular purposes. This creative, mental energy is akin to that recognized by religions since early time as World Soul, Brahman, or Holy Spirit. We do not know how it came into existence; it is sufficient to know that its existence is a hypothesis required to explain much of the activity in the phenomenal world.

4. The moral faculty enables mankind to recognize ethical values which can transcend the selfishness and viciousness of individuals and competing groups. Religions have from earliest times produced moral codes, commandments and rituals aimed at ensuring the good conduct of the cultural groups which they serve, and claiming divine authority they demand loyalty and obedience. These codes have come into being over thousands of years and reflect the historical times and cultures which gave rise to them. Like incompatible dogmatic religious beliefs, each claiming to be uniquely true, they are divisive, and while some of their edicts remain perennially valid, others become dated. They are no longer adequate to serve the world's multi-cultural and multi-faith population.

5. We now see that we are at a turning point in the world's evolution. If we continue on the present path of unrestrained material consumerism and expansion we are heading for environmental catastrophe and conflict between nations competing for the world's resources. At the same time man's spiritual energy has hitherto been concentrated on the divisive mythical beliefs of the old religions rather than the spiritual problems of the world we live in. The time has come therefore to change direction and identify spiritual values which can be accepted by all peoples of the world, and can provide beacons to guide the conduct of humanity, offering hope of unity and cohesion in place of divisiveness and conflict. These values have been outlined on pages 77–8 above. They do not depend for their validity on external supernatural authority, for ethical values carry their own ultimate authority in this world.

6. Faith in these global values will then provide an end in itself. The final destiny of life on our planet is unknown. Moral and aesthetic values will be re-evaluated from age to age, through democratic free expression of the spirit based on conscience, creativity, reason and wisdom, combined with knowledge. What man's understanding of these values will be thousands of years hence we can scarcely even guess. But by identifying from age to age the values by which the world can coexist in peace, the foundation is laid on which man can continue to explore the wonders of the natural world and the universe, and the mind/spirit which he shares with them.

7. When we consider the whole field of metaphysical beliefs and morality, two factors stand out as having the potential to provide unity and cohesion throughout the world in place of divisiveness:
   The universal consciousness which pervades the living world. It is this which, transcending the individual self, enables Homo sapiens to recognize that he is part of an interdependent living organism.
   The spiritual values which can be recognized and accepted by all cultures as common to humankind, and can provide the basis of a world ethic.
Together they form the foundation of a global philosophy which, given time, can bridge the divisions – religious, societal, economic and nationalistic – between the diverse cultures of mankind, and can offer hope for a future in which the peoples of the world can live in peaceful coexistence and in harmony with the environment.

8. With the advance of science, technology and industrial development, material and economic values have tended to submerge the spiritual. This can be seen at every level: in individuals selfishness and aggression tend to predominate, regardless of the harm done to others; local and regional institutions, corporate bodies and other communities tend to seek material advantage without regard

to wider social and national interests; and nation states seek to promote their narrow political interests, irrespective of whether this will increase the gap between developed and developing countries, jeopardize the well-being of future generations, or threaten other species and the environment. The time has come to redress the balance. This does not imply rejecting scientific knowledge and reason; it means that short-term material values need to be reassessed, and where necessary replaced, in the light of the universal spiritual values which are of ultimate importance for sustaining future life and development on earth.

9. Responsibility for how he lives his life, within the physical and other limitations placed on him, ultimately rests with each individual; he alone determines his moral values and decides how to base his conduct on them. In forming his ideals and goals each person can build on the moral teaching inherited from the past; he can draw not only on his own mental and spiritual resources, but combine with like-minded people dedicated to furthering their common aims. It is through education, the emergence of spiritual movements, and cultural interchange that attitudes and opinion can be changed throughout the world, so that the spiritual dimension can be brought into balance with the material. A world ethic can then provide the basis for a harmonious future for the planet instead of continued divisiveness and conflict.

10. As the spiritual life of the world is transformed in our pluralistic, multifaith world, the essential conditions are freedom of thought and expression, and tolerance. 'Love is wise, hatred is foolish. In this world, which is getting more and more closely connected, we have to learn to tolerate one another, and we have to learn to live together in that way. If we are to live together and not die together we must learn a kind of charity and a kind of tolerance which is absolutely vital to the continuation of human life on this planet.' But the exercise of tolerance presents its own problems, for tolerance should not be confused with indifference. When bigotry, fanaticism, fundamentalism and hatred conflict with the global values which are our goal, we should not be slow to condemn them and oppose the harm they do. We should work to see that they are superseded by the values which have the potential to unite the world instead of dividing it.

Bibliography

ALMOND, BRENDA, AND HARMAN, BRIAN (eds), *Values* (Humanities Press, 1988).
AYER, A. J., *Language, Truth and Logic* (Victor Gollancz, 1938).
AYER, A.J., *Ludwig Wittgenstein* (Penguin Books, 1985).
BRADLEY, IAN, *God is Green* (Darton, Longman and Todd, 1990).
CAMPBELL, JOSEPH, *The Hero with a Thousand Faces* (Princeton University Press, 1968).
CAPRA, FRITJOF, *The Tao of Physics* (Fontana Books, 1983).
CHARDIN, TEILHARD DE, *The Phenomenon of Man* (Collins, 1959).
COTTERILL, RODNEY, *No Ghost in the Machine* (Heinemann, 1989).
CUPITT, DON, *The Sea of Faith* (BBC, 1984).
GOULD, STEPHEN J., *Ever Since Darwin* (W. W. Norton & Co., 1979).
GOULD, STEPHEN J., *Wonderful Life: The Burgess Shale* (Hutchinson, 1990).
HAWKING, STEPHEN, *A Brief History of Time* (Bantam Press, 1988).
HOYLE, FRED, *The Intelligent Universe* (Michael Joseph, 1983).
INGLIS, BRIAN, *Natural and Supernatural. A History of the Paranormal* (Hodder and Stoughton, 1977).
KOESTLER, ARTHUR, *The Ghost in the Machine* (Pan Books, 1975).
LEWIS, CLARENCE I, *Mind and the World Order* (Charles Scribner's Sons, 1929).
LEWIS, C.S., *The Four Loves* (Geoffrey Bles, 1960).
MARKLEY AND HARMAN (eds), *Changing Images of Man* (Pergamon Press, 1982).
MASLOW, ABRAHAM, *The Psychology of Science* (New York, 1966).
MURRAY, GILBERT, *The Stoic Philosophy* (Conway Memorial Lecture, 1913).
PEACOCK, ROY E., *A Brief History of Eternity* (Monarch, 1989).
PEACOCKE, A. R., *Creation and the World of Science* (Clarendon Press, 1979).
PEDLAR, KIT, *Mind over Matter. A scientist's view of the paranormal* (Eyre Methuen Ltd, 1981).
PLATO, *The Republic* (translated A.D. Lindsay; J.M. Dent & Sons, 1929).
ROBINSON, JOHN A.T., *Honest to God* (Fontana Books, 1965).
RUSSELL, BERTRAND, *Power. A new social analysis* (Allen and Unwin, 1938).
RUSSELL, BERTRAND, *Problems of Philosophy* (Thornton Butterworth, 1936).
SAGAN, CARL, *The Dragons of Eden* (Hodder and Stoughton, 1977).
SCHUMACHER, E. F., *A Guide for the Perplexed* (Jonathan Cape, 1977).
STACE, W T., *Mysticism and Philosophy* (London, 1961).
UNITED NATIONS, *Our Common Future* (Report of the World Commission on Environment and Development, 1987).
WILSON, A N., *Against Religion* (Chatto and Windus, 1991).
WITTGENSTEIN, LUDWIG, *Tractatus Logico-Philosophicus* (Kegan Paul, 1933).
ZAEHNER, R C. (ed.), *The Concise Encyclopedia of Living Faiths* (Hutchinson, 1977).